UNDERSTANDING CLIMATE CHANGE

GRADES 7–12

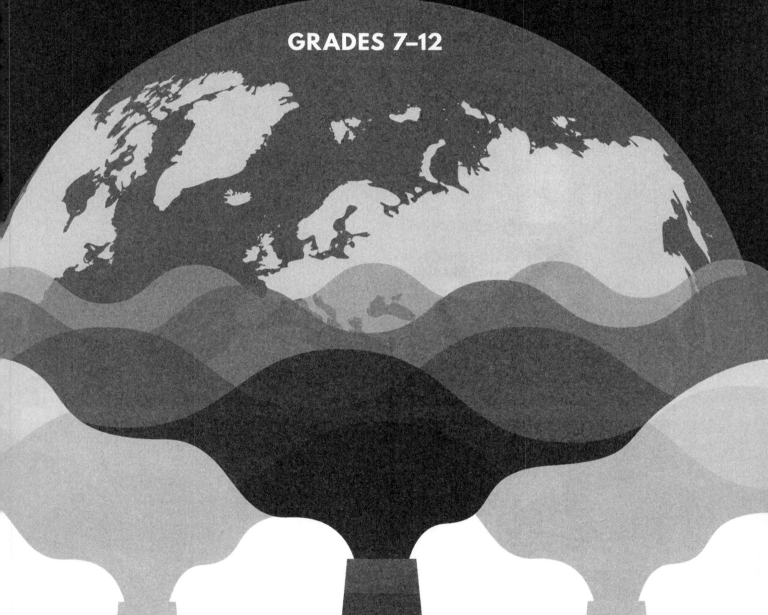

UNDERSTANDING
CLIMATE CHANGE

GRADES 7–12

LAURA TUCKER
LOIS SHERWOOD

NSTApress
National Science Teachers Association
Arlington, VA

National Science Teachers Association

Claire Reinburg, Director
Rachel Ledbetter, Managing Editor
Andrea Silen, Associate Editor
Jennifer Thompson, Associate Editor
Donna Yudkin, Book Acquisitions Manager

ART AND DESIGN

Will Thomas Jr., Director
Himabindu Bichali, Graphic Designer, cover design
Capital Communications LLC, interior design

PRINTING AND PRODUCTION
Catherine Lorrain, Director

NATIONAL SCIENCE TEACHERS ASSOCIATION
David L. Evans, Executive Director

1840 Wilson Blvd., Arlington, VA 22201
www.nsta.org/store
For customer service inquiries, please call 800-277-5300.

NSTA is committed to publishing material that promotes the best in inquiry-based science education. However, conditions of actual use may vary, and the safety procedures and practices described in this book are intended to serve only as a guide. Additional precautionary measures may be required. NSTA and the authors do not warrant or represent that the procedures and practices in this book meet any safety code or standard of federal, state, or local regulations. NSTA and the authors disclaim any liability for personal injury or damage to property arising out of or relating to the use of this book, including any of the recommendations, instructions, or materials contained therein.

PERMISSIONS

Library of Congress Cataloging-in-Publication Data
Names: Tucker, Laura, 1956- author. | Sherwood, Lois, 1950- author.
Title: Understanding climate change : grades 7-12 / by Laura Tucker and Lois Sherwood.
Description: Arlington, VA : National Science Teachers Association, 2019.
Identifiers: LCCN 2018052213 (print) | LCCN 2019001837 (ebook) | ISBN 9781681406336 (e-book) | ISBN 9781681406329 (print)
Subjects: LCSH: Climatic changes--Study and teaching (Secondary)--Activity programs.
Classification: LCC QC903 (ebook) | LCC QC903 .T83 2019 (print) | DDC 363.738/74071--dc23
LC record available at *https://lccn.loc.gov/2018052213*

Contents

Contents

Foreword

In the autumn of 2018, the National Science Teachers Association (NSTA) published a powerful position statement on its website about teaching climate change. And at the end of the year, *New York Times* writer David Leonhardt wrote, "There were more obvious big news stories than climate change in 2018. But there weren't any more important stories, in my view. That's why it is my choice for the top story of the year. It's the one most likely to affect the lives of future generations." Among all areas of science, climate science may well be the most critical for all citizens to understand. While the science of climate change is well understood by scientists, everyone will be affected by its consequences and everyone has a stake in how we respond.

Beyond its societal importance, climate science has a nearly unique pedagogical position as an inherently multidisciplinary, practical subject. Historical observations of weather and climate take into account everything from the physics of incoming and outgoing radiation balance to the chemistry of gases that absorb infrared radiation to the biology of photosynthesis and respiration applied to the land, ocean, and atmosphere of our planet. What's more, the subject is accessible to all ages: *A Framework for K–12 Science Education* articulates grade band endpoints for students from 2nd through 12th grade (NRC 2012, p. 188).

While many organizations and agencies have published materials to help educators teach about climate science, few books provide assistance for teachers in covering the scope of climate science with special attention to humanity's role. Laura Tucker and Lois Sherwood have set out to do just that with conscious attention to the three-dimensional teaching and learning called for in the *Framework*. Recognizing that many teachers have not had specific training in climate science, the authors provide useful summaries of the underlying science. The sections on the greenhouse effect, scientifically evaluating data, and conducting research on climate change topics are particularly helpful in recognizing the multidisciplinary aspects of climate science.

The response to climate change is not itself a scientific subject. Tucker and Sherwood are very clear that there is no debate about climate science. When considering what actions to take in the face of a changing climate system, we move from the scientific to the social or economic or political, where we need science to inform our arguments and decisions. The boundary between science as a means of understanding the natural world and the consequences of that understanding is often poorly defined. The NSTA position statement emphasizes the "science side" but recognizes the implications for society, as well. The authors have made a real contribution in this area by providing structured suggestions that encourage students to use the science they have learned in considering the effects of human activity. By making this connection, students have the best chance to use science to positively "affect the lives of future generations."

—David L. Evans, PhD
Executive Director
National Science Teachers Association

References

Leonhardt, D. *New York Times*. 2018. The Most Important Story of 2018. December 31.

National Research Council (NRC). 2012. *A framework for K–12 science education: Practices, crosscutting concepts, and core ideas.* Washington, DC: National Academies Press.

National Science Teachers Association (NSTA). 2018. *NSTA Position Statement: The Teaching of Climate Science.*

About the Authors

Laura Tucker has been a science educator for more than 40 years. Initially educated as a wildlife biologist, she found her passion teaching students in the outdoors. In 1979, she founded a nonprofit educational organization called Exploring New Horizons. It was designed to provide a comprehensive outdoor environmental science program for K–8 grade students and a summer camp program for children ranging from age 9 to 18. During her tenure at the organization, she helped develop a variety of programs that combined environmental science curricula (redwood, coastal, and Sierra Nevada natural history and ecology, marine biology, botany, zoology, geology, and astronomy) with music, dance, drama, art, and team building. The programs blended the teaching skills and talents of staff naturalists with those of classroom teachers to facilitate the incorporation of the outdoor school experience into the classroom. Approximately 60,000 students attended the programs while Laura was the executive director. Exploring New Horizons continues to this day, serving about 6,000 students per year on three campuses in California's Santa Cruz Mountains.

In 1992, Laura became the professional development coordinator for Great Explorations in Math and Science (GEMS), a nationally acclaimed resource for activity-based science and mathematics at the Lawrence Hall of Science at the University of California, Berkeley. While at GEMS, she worked with a variety of educators, including preservice teachers; classroom teachers; district, regional, and state curriculum coordinators; university faculty; and nonformal educators from museums, zoos, and nature centers. She was a leader in establishing the GEMS Network, which included approximately 72 sites and centers around the United States and 11 international locations. Laura served as a curriculum developer and reviewer for many GEMS publications, including *Aquatic Habitats* (Barrett and Willard 1998), *Dry Ice Investigations* (Barber, Beals, and Bergman 1999), *River Cutters* (Sneider and Barrett 1999), and *Schoolyard Ecology* (Barrett and Willard 2001) teacher guides. She also worked on handbooks that support the implementation of GEMS units and other programs.

Laura has focused a great deal of her energy on climate education. In 2012, she was selected as a Climate Reality Project presenter and joined former vice president Al Gore and 1,000 other educators from 59 countries for three days of intensive training. She is an NOAA Climate Steward as well as a team member of the Climate Change Environmental Education Project-Based Online Learning Community Alliance in partnership with Cornell University, the North American Alliance for Environmental Education, and the EECapacity Project. She serves as a mentor with Students for Sustainability, a group from Port Townsend High School in Port Townsend, Washington, that is taking action to mitigate climate change at their school, in their community, in their state, and at the national level. Laura also serves on the Jefferson County/City of Port Townsend Climate Action Committee and chairs the L2020 Climate Action Outreach Group. In December 2015, she attended the 21st Conference of the Parties in Paris where she conducted live interviews between young climate scientists and activists and students in her community.

Currently, Laura wears two hats. She is the waste reduction education coordinator for Jefferson County, Washington, teaching the community to reduce, reuse, and recycle. She is also a consultant, providing custom professional development for formal and informal educational programs in standards-based environmental and STEM (science, technology, engineering, and mathematics) education.

When Laura is not working to address the climate crisis, she is recharging her batteries by hiking, swimming, sea kayaking, soaking in hot springs, and enjoying the beauty of our natural world with her husband and their English bulldog, Yogi.

Lois Sherwood originally trained in zoology at Washington State University and worked in the medical profession after graduating. But after a chance visit to a high school classroom, she realized that teaching was her calling.

Lois began her teaching career at the SEA Discovery Center (formerly the Poulsbo Marine Science Center) and then moved to Port Townsend High School. During her career, she has taught health, marine biology, oceanography, physical science, math, biology, and integrated science.

While teaching, she earned a master's degree in science, also through Washington State University. As she deepened her understanding of science, she became interested in constructivist teaching and allowing students to learn through engaging in science practices. She refined this skill through inquiry training with the Exploratorium museum in San Francisco, California. Her training later led to codeveloping inquiry workshops, which she taught for several summers through the Port Townsend Marine Science Center.

Lois also served as a district teacher leader and as a Teacher on Special Assignment (TOSA) with the North Cascade and Olympic Science Partnership, which was a five-year science leadership project sponsored by Western Washington University and funded by a National Science Foundation grant. Along with a cohort of five other TOSAs, she promoted teacher leadership and best teaching practices in a five-county region of northwestern Washington.

With the cohort of TOSAs, she designed and led monthly regional workshops in addition to working with partner teachers in their classrooms.

Lois received National Board Certification in Teaching in 2007 and was recertified in 2017. For her work in the classroom, she received the Amgen Excellence in Science Teaching award in 2011 and was a finalist for the Presidential Award for Excellence in Mathematics and Science Teaching in both 2013 and 2015.

Beyond the classroom, Lois served as a regional representative with the Washington Science Teachers Association (WSTA) from 2007 to 2016. In 2011, she cochaired a joint statewide teachers conference with WSTA and the Environmental Education Association of Washington. In 2016, she was appointed as professional development coordinator for WSTA. In this role, she has facilitated the design and presentation of professional development offerings for the organization.

Connecting science to student involvement motivated Lois to facilitate a variety of student-led environmental clubs over the course of her career. Although the focus of the clubs has evolved based on student interest, the goal has always been to promote environmental and social justice locally, regionally, and nationally.

Lois's passion for science education is driven by a personal passion to understand and experience the natural world. This also fuels her hobbies, which include bird watching, beach exploration, kayaking, hiking, biking, and running.

References

Barber, J., K. Beals, and L. Bergman. 1999. *Dry ice investigations.* Berkeley, CA: Great Explorations in Math and Science.

Barrett, K., and C. Willard. 1998. *Aquatic habitats: Exploring desktop ponds.* Berkeley, CA: Great Explorations in Math and Science.

Barrett, K., and C. Willard. 2001. *Schoolyard ecology.* Berkeley, CA: Great Explorations in Math and Science.

Sneider, C., and K. Barrett. 1999. *River cutters.* Berkeley, CA: Great Explorations in Math and Science.

Acknowledgments

My endless appreciation goes out to Lois Sherwood, my exceptional coauthor, who brings her years of high school classroom experience and extensive knowledge of the *Next Generation Science Standards* (*NGSS*) to take this revised edition to a new level of excellence and effectiveness.

I am also grateful to Lois for allowing me to use her four 10th-grade classes to field-test the first edition of this book in 2011. After completing this unit, a number of students formed their own group—Students for Sustainability—and went on to make huge changes in their school and in their community. They even took *public transportation* from Washington state to Washington, D.C., to lobby for climate action—6,000 miles round-trip. They are featured as Climate Change Agents in Session 8.

This book would not have been possible without spending 20 years working with the brilliant and talented curriculum developers and staff of GEMS (Great Explorations in Math and Science) at the University of California, Berkeley's Lawrence Hall of Science. Their years of experience and keen insight into creating effective, teacher-friendly curricula have taught me well. They were on the cutting edge of climate change curricula with the book *Global Warming and the Greenhouse Effect*, written in 1990.

GEMS was the inspiration for *Understanding Climate Change*. Sessions 2 and 3 are adapted and modified from *Global Warming and the Greenhouse Effect*, copyrighted by The Regents of the University of California and used here with permission. Other sessions are partly inspired by the activities in the GEMS guide but are substantially revised, rewritten, and updated by the authors.

The scientific discourse circle in Session 4 is inspired by similar student-group activities in the *Seeds of Science/Roots of Reading* curriculum by GEMS, copyrighted by The Regents of the University of California and used here with permission.

Special thanks to the Climate Reality Project. Their extraordinarily talented staff conducts trainings around the globe and provides support for more than 15,000 Climate Reality Leaders, ranging from 12 to 86 years in age. I was honored to be included in their 2012 training in San Francisco that brought more than 1,000 leaders from 59 countries together for three days of information,

Acknowledgments

empowerment, and hope. We went out into our corners of the world to share the science of the climate crisis and provide a vision for how we can solve it.

I particularly want to thank all of the wonderful Climate Change Agents who so kindly agreed to be interviewed for this book and give a face to those working on the front lines for climate change solutions:

- James Balog
- Dr. Robert Bindschadler
- Dr. Shallin Busch
- Kate Chadwick
- Eliza Dawson
- Dr. Ziv Hameiri
- Dahr Jamail
- Rayan Krishnan
- Dr. Heidi Roop
- Ewan Shortess

My deepest gratitude goes to the exceptional staff at NSTA Press. To Claire Reinburg, my sincere appreciation for accepting my manuscript and moving it forward. To my phenomenal editors, Rachel Ledbetter and Andrea Silen, I am in awe of your ability to make sense of every word, every page, and every graphic. Thank you for weaving this all together so beautifully.

—Laura Tucker

Information for Teaching This Unit

Teaching With Fidelity for Student Comprehension

This unit has been designed with the Learning Cycle (Atkin and Karplus 1962) and the BSCS 5E Instructional Model in mind. In order to bring students along the continuum from awareness to comprehension, it is critical that the entire unit be taught with fidelity. It takes time to cover a topic as complex as climate change. It is recommended that this unit be taught once during grades 7–12. Of course, older students will have a greater ability to understand some topics, and different standards apply to middle and high school. Decide where this unit fits best in your school or district's scope and sequence to maximize the learning opportunities while building solid comprehension. Teachers are always encouraged to personalize their curriculum but should do so with a broad view, ensuring that key components are not left out for time's sake. It is recommended that teachers follow the suggested time frame so that complex topics are not short-changed and students are allowed ample time to engage, process, and reflect.

Conceptual Flow of the Unit

The unit has been strategically structured to engage student interest and build a conceptual foundation without overloading learners. It then provides students with the scaffolding to conduct their own research and draw their own conclusions about climate change. These sessions are linked and sequenced in such a way that students are able to build on concepts in order to better understand what is to come next, providing a constructivist model of learning.

Session 1 begins with a networking activity to give students a brief introduction to the Climate Change Agents they will meet throughout the unit. Then they share what they have heard about climate change, not what they know, in order to allow unfettered information to flow freely and provide the teacher with an understanding of student misperceptions. Questions are generated that serve to drive the entire unit as students uncover factual information supported by evidence.

In **Session 2,** students discover some sources of CO_2 in the atmosphere and compare their relative concentrations. The primary objective, in addition to being an interesting lab, is to show students that there is a considerable amount of CO_2 in car exhaust. This sets the stage for students to understand the role of fossil fuel combustion in warming our planet.

Session 3 uses two types of models to help students understand the key concept of the greenhouse effect and the role that CO_2 plays. Students then analyze data showing sources and quantities of greenhouse gases produced in the United States and worldwide.

The need to evaluate accurate information in a timely and effective way is addressed in **Session 4**, which concludes with students considering why common misconceptions might be held by some people. Emphasis on the need to support their claims with evidence is woven throughout this session.

Having gained a solid foundation for conducting research, students begin **Session 5** by forming research groups and selecting climate change–related topics. Students in each group investigate their chosen topic in order to develop a deep understanding of it. Teachers may choose for students to develop independent investigations to model their topic, as well. Groups then synthesize all of their information into presentations that they share in **Session 6** at a mock climate change conference, just as scientist share their findings in real life.

Session 7 has students reflect on the ripple effects of the topic they researched. For example, they might consider the initial effects of a three-foot rise in sea level for a coastal city. Then they consider the secondary effects and also list the human, environmental, and economic effects, which gives them a broader understanding of predicted changes resulting from a warming planet.

In **Session 8**, students use this same ripple effect model to look at the positive measures being taken to solve the climate crisis by local, state, national, and international agencies, corporations, and institutions. This lays the groundwork for **Session 9**, which brings together all they have learned in the unit to inform how they will make changes in their own lives and communities.

Structure of the Unit

Each session is constructed in a similar fashion with most—if not all—of the following components (not always in the same order):

1. **Introduction:** This is a brief overview of the session, providing the focus for the teacher.

2. **Objectives:** These are clear measures to define the skills and knowledge acquired by the students during the session.

3. **What You Need:** This includes detailed material lists divided by what you need for the class, what's needed for each group of students, and what each individual student needs.

4. **Preparation:** This provides instructions for how to prepare prior to and on the day of instruction, including what student handouts are necessary.

5. **Begin!:** This is the signal for the teacher to begin instruction after reading the preparatory information.

6. **Reviewing Statements and Questions About Climate Change:** In the first session, students record what they have heard about climate change in their notebooks. They also brainstorm and post questions that they have about climate change. At the end of each subsequent session, students review their notebooks and the questions as a way to measure their learning. They decide if any of the ideas listed in their notebooks can be put forth as a statement that is either accurate or inaccurate, based on evidence. They may also determine that new knowledge answers a question or leads to new questions. This is intended to be a dynamic process where students continually reflect and re-evaluate their learning. It serves as a critical step in cementing the conceptual knowledge that students gain in each session.

7. **Extending the Session:** This provides opportunities for students to delve deeper into the topic covered in the session or to challenge more advanced students.

8. **Extras Page:** All materials used by students (data sets, rubrics, worksheets, Climate Change Agent interviews, etc.) can be found on the web page *www.nsta.org/climatechange*. This allows the data sets to be updated frequently, keeping them current. Storing the materials online also allows for the production of higher-quality student copies.

9. **Climate Change Agent Interview:** In each session, students read at least one interview with someone who is working to address the issue of climate change. The interviewees, who are referred to as Climate Change Agents, range from students to scientists. Each is asked the same three questions:

   • Describe your work as a Climate Change Agent.

   • Describe your pathway to becoming a Climate Change Agent.

   • What do you think students should know about climate change?

10. **Background for Teachers:** This includes the science and pedagogy that support a deeper understanding of the content and process of each session. **The information is not to be read to the students**; rather,

it is designed to support teachers who may not have a background in this topic.

11. **Assessment Opportunities:** A variety of assessment modalities are provided to assist the teacher in determining the students' progress and depth of understanding. Student notebooks are designed to be used as assessment tools throughout the entire unit and as a final evaluation on student understanding and progress.

12. **Resources:** Additional information for both students and teachers is provided to support both the research component that is woven through the unit and enrichment opportunities.

Safety Considerations for Hands-on Activities

Teachers must address potential safety issues with engineering controls (ventilation, eyewash stations, etc.), administrative procedures/safety operating procedures, and appropriate personal protective equipment (indirectly vented chemical splash safety goggles or safety glasses meeting ANSI/ISEA Z87.1 D3 standard, chemical-resistant nitrile gloves, etc.). Teachers can make it safer for students and themselves by adopting, implementing, and enforcing legal safety standards and better professional safety practices in the science classroom and laboratory. Before undertaking any science activity or investigation, teachers should do a hazards analysis and risk assessment and then perform safety actions to ensure a safer teaching/learning experience. Remember that personal protective equipment is to be worn during the setup, hands-on, and takedown segments of the activity.

Always provide safety training and demonstrate proper use of hand tools, lab equipment, and personal protective equipment before having a student undertake any hands-on activities. Also provide follow-up safety reminders during each activity.

Throughout this book, safety notes are provided for classroom/laboratory activities. Teachers should also review and follow local policies and protocols used within their school district and/or school (e.g., chemical hygiene plan, Board of Education safety policies).

Additional standard operating procedures are provided by the National Science Teachers Association (NSTA). Visit NSTA's Safety in the Science Classroom web page (*www.nsta.org/safety*), which includes Safety Acknowledgment Forms for elementary, middle, and high school grade levels. Students should be required to review the document or one similar to it under the direction of the

teacher. Both students and their parents/guardians should then sign the document acknowledging the procedures that must be followed for a safer working/learning experience in the laboratory.

The Council of State Science Supervisors (CSSS) provides information about classroom science safety, including a safety checklist for science classrooms. See the CSSS website at *www.csss-science.org/safety.shtml* to access this information and get links to other safety-related resources.

Disclaimer: The safety precautions of each activity are based in part on use of the recommended materials and instructions, legal safety standards, and better professional practices. Selection of alternative materials or procedures for these activities may jeopardize the level of safety and therefore is at the user's own risk.

Unit Design With Standards in Mind

To meet today's standards, three-dimensional learning is integrated throughout the unit with specific references to performance expectations, crosscutting concepts, science and engineering practices, and disciplinary core ideas in the *Next Generation Science Standards* (*NGSS;* NGSS Lead States 2013). The concept of climate change provides exceptional opportunities for addressing the standards. The cross-curricular nature of the topic weaves in STEM (science, technology, engineering, and math), plus social studies, history, and current world problems. A correlation to the standards is provided for you in Table 1 (see pp. 12–13).

Background Information

Necessary Concepts

The following are concepts that are important for students and teachers to know before starting the unit.

For Students

It is recommended that students have some background knowledge of the following topics before beginning this unit in order to understand the concepts they will be required to learn as they move through the unit:

- Photosynthesis
- The structure and properties of matter (atoms, molecules, solids, liquids, gases)

- The carbon cycle (the basics of how carbon moves through earth, air, water, and living things by way of photosynthesis, respiration, and decomposition)

For Teachers

Current Carbon Versus Sequestered Carbon

Part of the debate about the amount of carbon dioxide in the atmosphere includes arguments from those who rationalize that CO_2 is "natural," because living organisms breathe it out in respiration. Of course, all carbon dioxide is natural, but the level of CO_2 in our atmosphere has been a determining factor in our climate for more than a billion years. Paleoclimatologists have concluded that CO_2 levels were five times higher when dinosaurs roamed the Earth than they are now. The seas were approximately 100 feet higher than they are now, and most of the Earth had a tropical climate. Yet, it is dangerous to imply that the existence of a CO_2 level this high in the past means it is a natural process and, therefore, not harmful. During these times, Earth had no human inhabitants, much less our current population of more than 7 billion people—many of whom live along coastlines that are particularly vulnerable to sea level rise caused by a warming planet.

For the past 800,000 years, the CO_2 level in our atmosphere has not risen above 300 parts per million (ppm) and has averaged 280 ppm. In 1850, the amount of atmospheric CO_2 was 285 ppm. In 1910, it crossed the 300 ppm mark. As of this printing in 2019, it sits at 412 ppm.

During those 800,000 years, our planet has evolved the ecosystems we live in today, with a livable temperature in most places and adequate rainfall for plants to thrive. This level of CO_2 has kept the heat energy on our planet from driving the weather systems to extremes. The CO_2 in this carbon cycle is considered *current carbon,* or the amount that has existed for this 800,000-year period of balance in our ecosystems.

With the advent of the Industrial Revolution in the late 1800s, oil deposits that were *sequestered* (or buried) from 540–65 million years ago were brought up to the surface of the Earth and burned, releasing their stored CO_2 into the atmosphere. This additional CO_2 caused the greenhouse effect to increase by more than 30%, resulting in numerous outcomes from a warming planet. Students will research these lines of evidence of global warming in Session 5.

History of the Study of the Greenhouse Effect and Climate Change

The earliest recorded thoughts about the greenhouse effect go back to the 1820s in France when Jean Baptiste Joseph Fourier (1768–1830) calculated that an object the size of the Earth should not be as warm as it was, given its distance

from the Sun. He reasoned that something else must be affecting our planet's temperature. He theorized that the light coming from the Sun was able to pass through our atmosphere, but radiant heat coming from Sun-warmed surfaces must somehow be trapped.

In the 1860s, John Tyndall (1820–1893), an alpine naturalist and climber, was fascinated by his observations of evidence that an ice sheet once covered northern Europe. Tyndall considered a number of possible explanations but settled on an idea that there could be variations in the composition of the atmosphere. He experimented with heat-trapping gases like water vapor and carbon dioxide. Even though there is such a small amount of CO_2 in the atmosphere, he thought it still could have an effect.

In 1896, Swedish scientist Svante Arrhenius (1859–1927) concluded that if atmospheric CO_2 levels doubled to 560 ppm (from preindustrial levels of 280), then surface temperature levels would rise several degrees. Arrhenius and his colleague, Arvid Högbom (1857–1940), started to consider the amount of carbon dioxide emissions from factories and were surprised to find that man-made emission rates were very similar to those occurring in nature. At the rate that coal was being burned in the 1890s, they didn't see this as a problem since it would take thousands of years for the doubling to take place. They also thought the oceans could absorb most of those emissions, so there was no cause for concern. This was the first suggestion that the burning of fossil fuels for heat could add enough CO_2 to the atmosphere to make a difference. They didn't have a concept of how this increased heat could affect the overall climate of the planet—only how it might melt glaciers.

Throughout the early to mid-1900s, there was considerable discussion, and considerable doubt, that the increase in CO_2 would have much of an effect on the planet. Again, the ocean was considered to be an endless carbon dioxide sink.

With the advent of the atomic age in 1945, carbon isotopes were able to be identified and compared in tree rings and other carbon-fixing forms. Scientists were able to distinguish between "old" carbon from fossil fuels and "current" carbon from their present atmosphere. They were also able to discern what percentage of CO_2 in their present atmosphere came from burning fossil fuels and what was part of the current carbon system.

In the mid-1950s, researcher Charles David Keeling (1928–2005) wanted to accurately measure the amount of CO_2 in the atmosphere but felt that using monitors near population centers would give inaccurate results. He settled on positioning sensors at the 13,600-foot-high Mauna Loa Observatory in Hawaii. Sitting at this altitude in the middle of the Pacific Ocean in 1958, Keeling began his measurements. They now serve as an important baseline, and scientists have been constantly monitoring and updating atmospheric CO_2 concentrations ever since. At the onset of his research, Keeling noticed a steady increase in CO_2, one much higher than what would be expected if the oceans were absorbing 80% of the emissions as had been commonly accepted. Keeling's son Ralph continues his father's research to this day with the data from the Keeling Curve (a daily record of atmospheric carbon dioxide) that's used extensively by scientists.

The first published use of the term *global warming* appears to have been by the climatologist Wallace Broecker in a 1975 article in the journal *Science* that was titled "Climatic Change: Are We on the Brink of a Pronounced Global Warming?" It is quite remarkable that a prediction made in 1975 using a simple model of the climate system could so accurately match the observed global temperature change we are seeing today. It is a testament to the dominant effect of CO_2 and the fact that we have had a solid understanding of the fundamental workings of the Earth's climate for many decades.

In June 1988, *global warming* became a more popular term after NASA scientist Dr. James Hansen told the U.S. Congress that "global warming has reached a level such that we can ascribe with a high degree of confidence a cause-and-effect relationship between the greenhouse effect and the observed warming." After Dr. Hansen's retirement from NASA in April 2013 (following 46 years of government service), he took on a more active role in the political and legal efforts to limit greenhouse gases.

The term *climate change* dates at least as far back as 1939. A closely related term, *climatic change,* was once also common, as exemplified in the 1955 scientific article "The Carbon Dioxide Theory of Climatic Change" by Gilbert Plass. By 1970, the journal *Proceedings of the National Academy of Sciences* published a paper titled "Carbon Dioxide and Its Role in Climate Change." In 1988, when the world's major governments set up an advisory body of top scientists and other climate experts to review the scientific literature every few years, they named it the Intergovernmental Panel on Climate Change (IPCC). This group is working on its sixth climate assessment, drawing on the work of hundreds of scientists from all over the world to enable policymakers at all levels of government to make sound, evidence-based decisions. The IPCC shared the Nobel Peace Prize in 2007 with Al Gore for "their efforts to build up and disseminate greater knowledge about man-made climate change, and to lay the foundations for the measures that are needed to counteract such change."

The Difference Between Global Warming and Climate Change

We often hear people say, "It used to be called global warming, now it's called climate change. Make up your mind!"

These are two different concepts. *Global warming* refers only to the rising temperatures of the Earth's oceans, land, and atmosphere, whereas *climate change* includes warming *and* the effects of warming—such as more severe storms, melting glaciers, or more frequent droughts, which may also result in increasing

forest fires. As the planet warms, there is more heat energy to run the Earth's systems like the water cycle and air currents, which affect the jet stream.

It is important to distinguish between the normal warming patterns that have occurred over time and the current patterns we are observing due to the burning of fossil fuels. In Earth's history before the Industrial Revolution, the planet was warmer due to natural causes not related to human activity. These trends are part of the natural cycles described by the Serbian astrophysicist Milutin Milankovitch in the 1920s. He hypothesized that the small changes in the Earth's orbit, axial tilt, and "wobble" cause enough difference in the amount of sunlight falling on the Earth to influence the climate. Ice cores taken in Greenland and Antarctica prove his hypothesis by showing temperatures warmer than they are now, hitting peaks about every 100,000 years.

Currently, the term *climate change* can mean human-caused changes to climate or natural ones, such as ice ages. On the other hand, *global warming* generally refers to human-caused warming—warming due to the rapid increase in carbon dioxide and other greenhouse gases from humans burning coal, oil, and gas since the Industrial Revolution. Besides burning fossil fuels, humans can cause climate changes by emitting aerosol pollution—the tiny particles that reflect sunlight and cool the climate—into the atmosphere or by transforming the Earth's landscape (for instance, from carbon-storing forests to farmland).

Some scientists use the term *climate disruption* to delineate the difference between a mere change (for instance, a slight uptick in average winter temperatures) and the type of swings in the weather that we see as the Earth warms. New York won't just change to be more like Florida, for example; rather, its climate will be different from what is normal for New York, which can include severe rainstorms, snowstorms, droughts, and floods. True, the average temperature in most locations will get warmer as we put more CO_2 into the atmosphere. More important, the additional heat and moisture in our atmosphere from global warming will cause major disruptions to what we used to consider normal for a particular location.

A simple distinction is that as the planet warms, it causes the climate to change, resulting in a cause-and-effect relationship between global warming and climate change.

"Believing" in Climate Change

As science educators, it is critical that we don't use the phrase *believing in climate change*. The *Oxford Dictionary of English* defines a belief as "an acceptance that something exists or is true, especially one without proof." Our beliefs are personal and do not need to be proven to anyone.

The nature of science, on the other hand, is reproducible results. A concept in science is accepted if the same situation is repeated over and over again, with

evidence collected that shows the same results. Only after repeatedly collecting evidence can we say whether the results support or do not support our initial predictions. If a study is published, we might say that we agree with the results of that study or accept the results of that study. If there are lots of replications and variations that all say the same thing, we might call the results a "fact," such as the fact that water freezes at 32°F or 0°C. We can agree with those facts. When the research of multiple lines of evidence all lead to the same conclusion, we might start talking about a theory, such as evolution. In the case of climate change, there are countless reputable studies that show the Earth is warming, causing changes in our climate. We don't believe in climate change; we accept the results of tens of thousands of studies and papers by reputable scientists that show our Earth is warming, resulting in a changing climate.

Controversy Over Climate Change

As scientific thinkers, skepticism is important. The term *skeptic* comes from the Greek noun *skepsis,* which means "examination, inquiry, and consideration." Questioning in science goes back thousands of years. The Royal Society of London for Improving Natural Knowledge, commonly known as the Royal Society, was founded in November 1660 and is the oldest national scientific institution in the world. Its motto, *Nullius in verba* (Latin for "on the word of no one" or "take nobody's word for it"), challenged scientists to question information presented to them. This is solid scientific practice. Scientists rely on reproducible results, large numbers of reputable studies, and respectful discourse to come to conclusions.

This is why it is hard to fathom the controversy over climate change, which has been researched at length. Despite the overwhelming scientific evidence for global warming and the observed effects on our climate, it can still be a controversial topic.

Why is climate change and its effects still disputed? Perhaps the best explanation can be found in the age-old saying "Follow the money." James Hansen claims that in order to keep our planet from warming more than 2°C, a commonly accepted limit by scientists, we can only put another 565 gigatons of CO_2 into our atmosphere (Hansen et al. 2013). The Carbon Tracker Initiative, a team of London financial analysts, estimates that proven coal, oil, and gas reserves of fossil fuel companies and countries would put an additional 2,795 gigatons of CO_2 into the atmosphere if extracted and burned (Leaton 2011). That is five times the amount we can release to maintain a 2°C limit of warming. At today's market values, those reserves are estimated to be worth $27 trillion in U.S. dollars. That's 27 trillion reasons to burn as much fossil fuel as possible, if you're in the business of selling those reserves. It's also a $27 trillion loss if the reserves are kept in the ground. The controversy is essentially an economic issue and should not be a scientific or political one.

In August 2017, for example, two Harvard researchers published a paper assessing whether ExxonMobil Corporation has in the past misled the general public about climate change (Supran and Oreskes). They presented an empirical document-by-document analysis and comparison of 187 climate change communications from ExxonMobil, including peer-reviewed and non-peer-reviewed publications, internal company documents, and paid, editorial-style advertisements in *The New York Times*. The researchers looked at the company's views of climate change as real, human-caused, serious, and solvable. They stated in their conclusion that "the company's apparent acknowledgment of climate science and its implications seems dramatically at odds with basically its current business practice." Those 27 trillion reasons would most definitely affect stockholders for ExxonMobile and other fossil fuel companies. The actions taken by these companies might be best summed up by an Upton Sinclair quote: "It is difficult to get a man to understand something when his salary depends on his not understanding it."

In order to get results that stand up against self-interested parties and avoid undue skepticism, scientists must conduct their research with impunity and carefully examine all of the data, not just data that support their predictions. What's more, research funding can raise issues when the results of a study funded by a particular company or group are contrary to that company's or group's bottom line. Therefore, scientists must guard against the influence of funders.

The "Debate" Over Climate Change

It is probable that discussion of the presumed controversy over human-caused climate change will arise in the classroom during this unit. It is important to remember that 97% of scientists worldwide agree that the causes of our rapidly warming planet are directly related to the burning of fossil fuels. To allow a debate by "both sides of the argument" is akin to having a debate about geocentric versus heliocentric models of the solar system. Although a geocentric model was widely accepted in its day, overwhelming scientific evidence has shown that model to be inaccurate, and it has not been accepted for more than 200 years.

The same is true for climate change. As of this printing in 2019, the United States is the only country to withdraw from the Paris climate accord, which was signed by 195 countries. At the 21st Conference of the Parties (COP21) in 2015, every "party" (each country, plus the European Union) came together to set limits that would ensure a sustainable future for humans on Earth. Of all the parties at the conference, only two did not sign the agreement. One was Nicaragua, which claimed the limits did not go far enough. The other was Syria, which was steeped in a civil war. Since then, both Nicaragua and Syria have signed the agreement. In 2017, the United States committed to withdraw, taking effect in 2020. This should make any skeptic ask, "Why?" Every other country in the world agrees that climate change is real, is primarily caused by burning fossil fuels, and if unchecked could make Earth uninhabitable to most humans. They also agree there are solutions that need to be put in place as soon as possible.

Table 1: NGSS Correlations

Next Generation Science Standards	Session								
Performance Expectations	1	2	3	4	5	6	7	8	9
MS-PS4-2			**						
MS-ESS3-5				**			**		
HS-ESS2-2			**		**				
HS-ESS2-4			**	**					
HS-ESS3-5				**	**				
HS-ESS3-6							*		
HS-ETS1-3								*	*
HS-ETS1-4									*
Disciplinary Core Ideas	1	2	3	4	5	6	7	8	9
PS4.B			*						
ESS2.A			**		**				
ESS2.D			**	**			**		
ESS3.D			**	*	**				
ETS1.B								*	*

Continued

Table 1 (continued)

Next Generation Science Standards	Session								
Crosscutting Concepts	1	2	3	4	5	6	7	8	9
Patterns		*	*						
Cause and Effect		*	**	**	**		**	**	
Scale, Proportion, and Quantity			*	*	*				
Systems and System Models		*	**		**		**	**	*
Energy and Matter			*						
Structure and Function			*						
Stability and Change			**	**	*				
Science and Engineering Practices	1	2	3	4	5	6	7	8	9
Asking Questions and Defining Problems	**	*	**	**	(**)		*	*	
Developing and Using Models			**		(**)				**
Planning and Carrying Out Investigations					(**)				
Analyzing and Interpreting Data		**	**	**	**				
Using Mathematics and Computational Thinking				*	**				*
Constructing Explanations and Designing Solutions		*	**	**	**				*
Engaging in Argument From Evidence				*		**	*		
Obtaining, Evaluating, and Communicating Information			*	*	**	*	*	*	

** = addresses all elements of standard   (**) = standard is addressed in optional investigations

* = addresses most of standard but will require some additional teaching

References

Atkin, J. M., and R. Karplus. 1962. Discovery or invention? *The Science Teacher* 29 (5): 45–51.

Broecker, W. 1975. Climatic change: Are we on the brink of a pronounced global warming? *Science* 189 (4201): 460–463.

Do the Math. FAQs. *http://math.350.org/questions* (accessed November 3, 2018).

International Panel on Climate Change. History. IPCC. *www.ipcc.ch/organization/organization_history.shtml* (accessed November 3, 2018).

Leaton, J. 2011. *Unburnable carbon: Are the world's financial markets carrying a carbon bubble?* Carbon Tracker Initiative. *www.carbontracker.org/reports/carbon-bubble.*

Mason, J. 2013. The history of climate science. Skeptical Science. *https://skepticalscience.com/history-climate-science.html* (accessed November 3, 2018).

Muller, R. *New York Times.* 2012. The conversion of a climate-change skeptic. July 28. *www.nytimes.com/2012/07/30/opinion/the-conversion-of-a-climate-change-skeptic.html* (accessed November 3, 2018).

NASA. Global mean CO_2 mixing ratios (ppm): Observations. *https://data.giss.nasa.gov/modelforce/ghgases/Fig1A.ext.txt* (accessed November 3, 2018).

NASA. 2008. What's in a name? Global warming vs. climate change. *www.nasa.gov/topics/earth/features/climate_by_any_other_name.html* (accessed February 4, 2019).

NGSS Lead States. 2013. *Next Generation Science Standards: For states, by states.* Washington, DC: National Academies Press. *www.nextgenscience.org/next-generation-science-standards.*

NOAA. CO_2 at NOAA's Mauna Loa Observatory reaches new milestone: Tops 400 ppm. *www.esrl.noaa.gov/gmd/news/7074.html* (accessed November 3, 2018).

Norweigan Nobel Committee. 2007. Prize announcement. Nobel Media AB. *www.nobelprize.org/prizes/peace/2007/summary.*

Oxford Dictionaries. 2019. *Oxford Dictionary of English.* Oxford University Press. Continually updated at *https://en.oxforddictionaries.com.*

Supran, G., and N. Oreskes. 2017. Assessing ExxonMobil's climate change communications (1977–2014). *Environmental Research Letters* 12 (8): 084019.

Sustainable Innovation Forum. Find out more about COP21. *www.cop21paris.org* (accessed November 3, 2018).

SESSION 1

What Have You Heard About Climate Change?

Introduction

The goal of this session is to set the stage for student investigations into the complex topics of global warming and climate change. Students first participate in a networking activity to introduce them to the Climate Change Agents whose interviews are found throughout the guide. They then activate their prior knowledge through a brainstorm of what they have heard about climate change. In order to reveal student understanding and misconceptions about climate change, the term *heard* is used rather than the term *know*. This allows students to feel free to share anything they have heard without fear of being wrong while the teacher is able to uncover their misconceptions prior to further instruction. This process sets up thinking and communication practices that prepare the students for *Next Generation Science Standards* (*NGSS*) performance expectations that are developed in the subsequent sessions. After listing ideas they've heard about climate change, students review one another's lists and formulate questions that they have about climate change.

A formative assessment probe from *Uncovering Student Ideas in Earth and Environmental Science* by Page Keeley and Laura Tucker is included to assist teachers in assessing whether their students have a firm understanding of the difference between climate and weather.

Objectives

1. To encourage students to share their own information about what they have heard regarding climate change, which serves as a preassessment of their prior knowledge

2. To establish a base of common information that will be useful both for teaching and for evaluating the information covered in the unit

3. To assess what knowledge and misconceptions students hold about climate change

What You Need

Gather the following materials.

For the class:

- ☐ Package of sentence strips (usually 100 in a package)
- ☐ Roll of masking tape

For each group of four students:

- ☐ Marking pens
- ☐ 5 or more sentence strips from the class package

    Note: You can make your own sentence strips rather than buying them. Use butcher paper cut into 3 × 36 in. long strips, or if the sentence strips are too cumbersome, choose a shared document platform that students can access from digital devices.

For each student:

- ☐ Science notebook
- ☐ Networking card from Handout 1.1: Networking
- ☐ Copy of Handout 1.2: Climate Change Agent Interview
- ☐ Copy of assessment probe Are They Talking About Climate or Weather?

    Note: The handouts and probe are located on the Extras page: www.nsta.org/ climatechange.

Preparation

A Few Days Before the Class

1. Find a space on the wall for four columns of sentence strips.

2. Write the following column headings on four different-colored sentence strips:

   - Questions We Have About Climate Change

   - Accepted as Accurate—Supported by Evidence

   - Accepted as Inaccurate—Supported by Evidence

   - Needs More Information/Evidence/Research

   Table 1.1 provides an example of how the columns should look once you set them up on the wall during class.

Table 1.1: Example of Wall Columns			
Questions We Have About Climate Change	Accepted as Accurate— Supported by Evidence	Accepted as Inaccurate— Supported by Evidence	Need More Information/ Evidence/ Research
• question • question	• statement • statement	• statement • statement	• statement • statement

3. Decide how you will attach the sentence strips to the wall. A recommended method is to attach two long strips of masking tape to the wall for each column of sentence strips, sticky side out. Place the strips about two feet apart from each other. You may also use poster putty instead of tape.

4. Decide how you will manage the sentence strips for each class. You can roll them up after each period or compile them as a group, depending on how many classes you have.

   Note: *It is important that the statements in each category can be moved back and forth as new information is discovered. For example, a statement in the Accepted as Inaccurate column might be moved to the Needs More Information/Evidence/ Research section when additional information is revealed. Questions can be removed when they are answered.*

The Day Before the Class

1. Prepare Handout 1.1: Networking for the class. The five-page-long handout includes 10 different networking cards, with 2 cards per page. Each card features a quote from a different Climate Change Agent. Make enough copies so that each student in your class receives a card. So, for example, if you have 30 students in your class, make three copies of the handout. Cut each page in half after printing the copies to separate the cards. (Given that there are only 10 separate cards, some students will likely have the same interview.)

2. Make copies of Handout 1.2: Climate Change Agent Interview.

Begin!

Networking

1. Distribute one networking card to each student. Tell students that they will take on the identity of the person on their card and attend a "networking event" with many other very important people. Let students know that name-dropping is important at networking events.

2. Ask students to walk around the room and gather information from as many different people as possible. As they meet each person, they should exchange names and read the quotes on their cards to each other. Students then record the names and key ideas from each person they meet. If your school policy allows it, students may take pictures of each other's cards with their cell phones.

3. After about five minutes, have students return to their tables. Ask them to work with their table partners to compare their impressions of the people they met at the networking event. Once students have finished discussing their impressions, tell them that they will have an opportunity to read more about each of these people in the upcoming unit.

4. Ask students to discuss the following questions in groups and record their thinking in their notebooks:

   • What is the common thread between all of the people at the networking event?

   • Which person are you most interested in learning more about? Why?

   • What do you think you will be learning in the upcoming unit?

5. Give students an opportunity to share their answers with the class.

Drain Your Brain About Climate Change

1. Tell the class that you want to find out what they have *heard* about climate change.

   Note: *Use the term* heard *rather than* know *as it takes pressure off of students to have correct answers and prompts them to share more freely. This will allow you to be more effective in assessing your students' prior knowledge.*

2. Invite students to think about everything they have heard about climate change from television, the internet, social media, books, newspapers, or word of mouth. Anything they have heard is acceptable.

3. Tell students they each have a few minutes to make a list in their notebooks of everything that comes to mind. Each student should have at least four items in their list. Tell them not to worry about using whole sentences or whether or not they are certain of the information. They don't need to personally agree with what they have heard from a particular source, either. It is good to get all information and misinformation out where the students can discuss it.

4. Ask them if they have questions about what they are going to do, and then begin.

5. If some students run out of things to write, have them write questions they have about climate change.

Thought Swap

1. After a few minutes, or once most students are finished, have them stop writing.

2. Explain that it is time for a Thought Swap. Students will work in groups of three to four. The directions are as follows:

   a. Each student will read their list to the group. No one is allowed to interrupt the speaker or comment on the information that is shared at this time.

   b. After one person finishes sharing, the next person has a turn, stating only those things not covered by the first speaker. This continues until everyone has shared the content of their lists. Students may add information shared by their classmates to their own lists.

   c. After all group members have shared their information, they can ask one another questions and discuss what each person had to say. Any disagreements can be discussed at a later time.

3. While the group discussion is going on, post the following four sentence-strip headings to your wall columns or use a computer to project the categories:

- Questions We Have About Climate Change

- Accepted as Accurate—Supported by Evidence

- Accepted as Inaccurate—Supported by Evidence

- Needs More Information/Evidence/Research

  If sentence strips are being used, leave room under each heading for 15–20 statements to be posted.

4. Ask if there are any questions before moving on.

Sharing What We've Heard About Climate Change

1. After about 10 minutes (or when most groups are finished), stop the small-group discussions. Ask each group to choose two of the things that they have heard about climate change to share with the whole class. Tell groups that if what they decide to share is presented by another group first, then they will have to pick something else to share. Do not make any comments at this time about the accuracy or inaccuracy of the information.

2. After each group shares, ask if anything that they had heard about climate change seems to be contradictory. Mention that climate change is a complex topic, and misconceptions are very common. Students may find conflicting information as they study the topic further. However, the job of conscientious scientists is to back up any statement with solid evidence, and the students will all be acting like conscientious scientists throughout the unit. You may want to point out that what is accepted in science can change as new information is discovered or revealed, and what we thought to be accurate in the past may no longer be accurate today.

3. Point out the column headings you placed on the wall. Mention that at this point in the unit, students most likely will not have enough information to decide whether the ideas they have heard about climate change can be accepted as accurate or inaccurate, so you will leave those categories empty for the time being. As they go through the unit, students will be conducting investigations, comparing data, and performing their own research to determine the accuracy or inaccuracy of what they have heard about climate change. For now, they will keep all of their ideas written in their notebooks for future reference.

Note: Students may bring up the ozone hole, acid rain, or other environmental issues that they confuse with climate change. You may choose to ask them to eliminate those statements from their list or leave them in as part of the discussion.

Examples of What Students May Have Heard About Climate Change

"What I have heard about climate change …"

- If the climate changes, it will cause glaciers to melt, which will eventually raise the sea level.

- It's caused by ozone-layer thickening.

- CO_2 levels are increasing.

- Ocean temperatures are increasing.

- Climate change causes storms to grow larger.

- Climate change makes the world warmer.

- A portion of the United States does not recognize it as a problem.

- The polar ice caps are melting, raising ocean levels.

- Global warming is bad.

- Climate change is caused by greenhouse gases.

- Pollution is depleting the ozone layer.

- It is making our atmosphere thinner.

- It is making the world unstable.

- Burning fossil fuels causes greenhouse gases.

- Natural disasters are getting worse.

- It is thought to be more extreme now than in the past.

- Arctic species are going to be extinct in 20 years.

The Difference Between Weather and Climate

1. It is important that students have a clear understanding of the difference between the terms *weather* and *climate* before continuing with this unit. *Weather* refers to the conditions at one particular time and place. These conditions can change from hour to hour, day to day, and season to season. *Climate,* on the other hand, refers to the long-term average pattern of weather in a location. For example, we might say that the climate of South Florida is warm, moist, and sunny, although the weather on a particular day in this region could be quite different than that. Long-term data are needed to determine changes in climate.

2. To be sure students have a firm grasp of this foundational component of the study of climate change, administer the formative assessment probe Are They Talking About Climate or Weather? The probe can be found on the Extras page (*www.nsta.org/climatechange*). Answers and sample student responses to the open-ended question in the probe can be found in the Resources section (pp. 25–27) at the end of this session.

3. If students do not demonstrate a clear understanding of this fundamental concept, review reteaching suggestions in the Pedagogy section (pp. 24–25) of this session.

What We Don't Know About Climate Change

1. Ask students in each group to review their notes on what they have heard about climate change. From those notes, they will brainstorm questions they have about climate change. They should have at least one question per person in the group.

2. Students will post their questions in one of two ways:

   - Have students in each group write their questions on sentence strips. Then they can bring the strips up to the wall and post them in the Questions We Have About Climate Change column. As the questions are being posted, have students monitor the board so there are no duplicate questions.

   - If using Google Docs or another electronic posting medium, have students in each group post their questions in the document you have created. Be sure to monitor what the students are posting to avoid duplicates and/or inappropriate comments.

3. Once all of the questions are posted, have students take turns reading them aloud to the class.

Examples of Student Questions

"Questions that I have about climate change …"

- Why is it a big deal?
- Is there a scientific experiment on solving climate change?
- Will it cause any land to disappear?
- How can we stop it?
- How long will it take to get too hot to live?
- How long has it been going on?
- Will it cause the world to end?
- How does the release of animal methane affect the ozone?

- Why have we not fixed it when we have known about this for so long?
- Are our advances in science and technology helping or hurting climate change?
- What are the contributors?
- How do automobiles compare with factories in contributing to climate change?
- Is climate change speeding up or at a steady pace?

4. Ask students to raise their hands if they know the answers to any of the questions currently posted. Have them silently write their answers to these questions in their notebooks without discussion with other students. Explain that they will have a chance to review these questions and answers as the unit progresses.

Climate Change Agent Interview

1. Distribute Handout 1.2 to the class. It features the story of Eliza Dawson, who planned to row from California to Hawaii to draw attention to the issue of climate change. Ask students for their initial impressions of the interview. Take a few responses. Refrain from getting into long discussions at this time. This interview is designed to get students engaged in the topic of climate change and to introduce a young scientist who is so passionate that she and three friends planned to row more than 2,400 miles across the Pacific Ocean to study climate change.

2. Ask students for their impressions of Eliza's major setback in her quest to row to Hawaii and how she handled the setback. Do they agree or disagree with her choices? Again, this first interview is meant to give them a brief introduction to the issue of climate change and whet their appetite for future interviews.

Background for Teachers

Definitions

As you go through the session, you may find the following definitions from the Environmental Protection Agency helpful.

Climate: The average weather conditions in a particular location or region at a particular time of the year. Climate is usually measured over a period of 30 years or more.

Climate Change: A significant change in the Earth's climate. The Earth is currently getting warmer because people are adding heat-trapping greenhouse gases to the atmosphere. The term *global warming* refers to warmer temperatures, whereas *climate change* refers to the broader set of changes that go along with warmer temperatures, including changes in weather patterns, the oceans, ice and snow, and ecosystems around the world.

Global Warming: An increase in temperature near the surface of the Earth. Global warming has occurred in the distant past as the result of natural causes. However, the term is most often used to refer to recent and ongoing warming caused by people's activities. Global warming leads to a bigger set of changes referred to as *global climate change*.

Weather: The condition of the atmosphere at a particular place and time. Some familiar characteristics of the weather include wind, temperature, humidity, atmospheric pressure, cloudiness, and precipitation. Weather can change from hour to hour, day to day, and season to season.

Pedagogy

What Students Have Heard Versus What They Know

Teachers have often observed that students will remember inaccurate statements as true when they see them posted for the whole class to see. This is why students are instructed to put what they have *heard* in their notebooks and not where the whole class can read the statements. This helps individual students reflect on their own ideas and misconceptions and correct them as the unit goes on. Once a statement has been determined as accurate or inaccurate by the class based on evidence they have discovered, then the statement can be placed where it can be seen by the class. Teachers can monitor a specific student's conceptual path by reading their science notebook.

Addressing Misconceptions in Climate Versus Weather

The upcoming sessions will direct students to dig deeply into the data associated with climate change, research the lines of evidence supporting climate change, and engage in scholarly discourse. For this to have meaning, students need to have a sound understanding of the difference between climate and weather. Many students can recite the catchphrase that climate determines the clothes in your closet, and weather determines the clothes that you select to wear today. However, when asked to apply their understanding to real-world situations, students (and adults, too!) often lack deep understanding. To evaluate this, ask students to complete the formative assessment probe Are They Talking About Climate or Weather? Answers and sample student responses to the open-ended question in the probe appear at the end of the Resources section (pp. 26–27). As you review students' results, you may find that you need to do some reteaching.

Select the resources that you want to use for reteaching. You may select readings from the students' textbook or appropriate video shorts to share with your class. (Two short videos are included in the Resources section.) Alternatively you may give students an opportunity to research weather and climate online.

After showing a video or giving students time for personal research, ask them to record their own definitions of weather and climate in their notebooks.

Once they have recorded their own thinking, ask students to share their definitions with others in their group. When everyone has shared, each group will create one definition for the term *weather* and one definition for the term *climate*.

Have students share their group definitions with the class. Ask students to make any additions or corrections to their group definition, then create a class definition for the terms *climate* and *weather*. You may use the definitions provided in the Background for Teachers section (pp. 23–24) to guide student thinking.

Hand back students' formative assessment and allow them to make corrections to their original responses. Review responses with the class quickly. If you come to a response where there is disagreement, allow students to make the case for their selections. Using the class definitions, come to consensus for each item. This is a great place to give students an opportunity to engage in argumentation, which is both a science and engineering practice in the *NGSS* and a *Common Core State Standard* for both English language arts and math.

Resources

1. These videos help students understand the difference between weather and climate.

   - This video may be more appropriate for middle school students: *www.youtube.com/watch?v=YbAWny7FV3w.*

   - This video may be more appropriate for high school students: *www.youtube.com/watch?v=VHgyOa70Q7Y.*

2. The following skit by Deep Rogue Ram shows a weather reporter gone rogue when she is confronted by her coanchor, who is rather clueless on the issue of climate change: *www.youtube.com/watch?v=TmfcJP_ 0eMc&t=6s.*

3. Skeptical Science (*www.skepticalscience.com*) is an excellent resource with commonly accepted misconceptions about global warming and climate change along with corresponding scientific explanations. You can choose different levels of scientific explanations, from basic to

intermediate to advanced. DO NOT SHARE this information with students at this time. Students will explore common misconceptions in Session 4. This resource is listed to help you understand some of the common misconceptions your students will have heard.

4. National Oceanic and Atmospheric Administration (NOAA) offers a climate literacy guide called *Climate Literacy: The Essential Principles of Climate Sciences*. The guide presents a vision of a climate-literate society. Many scientists and educators collaborated to produce this guide, building on efforts to define climate literacy and identify the principles and concepts of climate science that should be included in K–12 curricula. You can find the guide here: *https://oceanservice.noaa.gov/education/literacy.html*.

5. The Yale Program on Climate Change Communication has been conducting some of the most comprehensive research on public opinion and behavior concerning climate change. In their studies, they found that 63% of Americans believe that global warming is happening, but many do not understand why. Only 8% of Americans would receive a letter grade of an A or B for their knowledge of global warming, 40% would receive a C or D, and 52% would get an F. The studies also found important gaps in knowledge and common misconceptions about climate change and the Earth system. These misconceptions lead some people to doubt that global warming is happening or that human activities are a major contributor, to misunderstand the causes and the solutions, and to be unaware of the risks. For more information, go here: *http://climatecommunication.yale.edu/publications/americans-knowledge-of-climate-change*.

6. For the Are They Talking About Climate or Weather? assessment probe, the best answer is that B, E, and H are statements related to climate. Answer choices A, C, D, and G are statements related to weather. Students have effectively argued that answer choice F can relate to both climate and weather. Sample responses to the open-ended question in the probe are listed here:

   • What had to do with clothing was W and the rest was C, but that did not work for some.

   • I chose climate for the time of year and weather for anything that had to do with weather.

   • If the weather was not usual, then I put C, and if it was, I put W.

   • I think climate hasn't happened; it might happen. Weather already happened, like it is raining.

- I put weather for the ones that state rain, drought. Climate is what makes the weather happen.

- It's climate if it's a change, and it's a W if it's something to do with how the outdoors affects your day.

- Rain is something that happens outside. That is climate. Weather is something you might have to change your clothes for.

- I think climate when I hear rain, because global warming melts icebergs, which means more water, and water evaporates quicker with more rain.

- My reasoning is that weather is what happens every day and is mostly unpredictable, and climate is more predictable and is what happens over a number of years.

- Weather is rain, snow, hail, sunshine, and wind, and climate is density. Density is like humidity or moisture in the air.

- Weather is random. Climate change is big news and at a certain place.

- Weather is the conditions such as raining, snowing, sunny, etc. Climate is conditions of the air, such as humid, muggy, dry, and others.

- If the sentence referred to a specific weather condition such as snow, rain, or temperature, I counted it as weather. If the sentence referred to natural disasters or did not mention a specific weather condition, I counted it as climate.

References

Bybee, R. et al. 2006. The BSCS 5E Instructional Model: Origins and effectiveness. Paper prepared for the Office of Science Education at NIH.

Environmental Protection Agency (EPA). 2017. A student's guide to global climate change: Glossary. *https://archive.epa.gov/climatechange/kids/glossary.html#w.*

Keeley, P., and L. Tucker. 2016. Are they talking about climate or weather? In *Uncovering student ideas in Earth and environmental science*, 74–76. Arlington, VA: NSTA Press.

SESSION 2

Sources of CO$_2$ in the Atmosphere

Introduction

Carbon dioxide (CO$_2$) is the primary gas that creates our "Goldilocks" planet—not too hot, not too cold, just right. It serves as a heat-trapping blanket in our atmosphere, providing a natural greenhouse effect that allows Earth to maintain an average temperature of 15°C (59°F). Without CO$_2$ in the atmosphere, heat would radiate back into space causing Earth to be a mostly frozen ball with an average temperature of −18°C (0°F).

Students may have heard that climate change is primarily a problem of too much carbon dioxide in the atmosphere. It has caused most of the warming we have measured since the mid-1700s and is caused primarily by the burning of fossil fuels such as coal, oil, and gas, or the cutting down and burning of forests. Currently, there is 30% more CO$_2$ in our atmosphere than at any time in the past 800,000 years.

Before students dive into the causes and effects of the increase in this critical greenhouse gas, they are provided with an opportunity to discover some of the sources of CO$_2$ in our atmosphere. Students conduct an investigation where they collect gas samples from the air, from their breath, from a gasoline-powered automobile, and from a pure source of CO$_2$. They compare the relative concentrations of CO$_2$ by bubbling their gas samples through bromothymol blue, an easy-to-obtain acid/base indicator. Their results show that humans produce a small amount of CO$_2$ through respiration as compared to the significant amount of CO$_2$ produced by a gasoline-powered automobile. This experiment provides some foundation for understanding the role that the burning of fossil fuels plays in our warming planet.

 **Session 2**

Objectives

1. To give students a concrete understanding of some sources of CO_2 and their relative concentrations in our atmosphere

2. To practice sampling and testing techniques like those used in scientific studies of the atmosphere

3. To draw out the distinction between "natural" and "industrial" sources of CO_2

What You Need

Gather the following materials.

For each class:

- ❑ 1 lb. boxes of baking soda (2 per class)
- ❑ 750 ml of diluted bromothymol blue (BTB) in a clear 1 L bottle or in a 1 gal. jug (You may need distilled or deionized water if making BTB from powder. See information on preparation and safety in Preparation section.)
- ❑ 1 gal. (3.8 L) jug of distilled white vinegar
- ❑ Extra balloons beyond what is needed per group, in case they pop
- ❑ Technology to project a web page for the class to view
- ❑ The current list of questions and statements on the wall OR access to the electronic documents with questions and statements
- ❑ Sentence strips for additional statements (if you're using the wall columns)
- ❑ Marking pens (if you're using the wall columns)

Shared with all classes:

- ❑ Air pump (either bicycle pump or balloon inflator)
- ❑ Car and driver, plus 1–2 adult supervisors
- ❑ Manila folder
- ❑ Whiteboard or chart paper and marker

For each group of four students, plus one extra set for your demonstration:

- ❑ Lab tray to contain all the materials (the 1 L glass bottle can be separate)
- ❑ Empty 1 L glass bottle with no rough edges around the top of the bottle (or the balloons will pop when placed over the mouth of the bottle)
- ❑ 200 ml or larger squirt bottle containing the dilute BTB solution
- ❑ 2 rolls of masking tape
- ❑ Teaspoon
- ❑ 100 ml graduated cylinder or measuring cup
- ❑ 8 oz. plastic cup labeled "Vinegar"
- ❑ 8 oz. plastic cup labeled "Baking Soda"
- ❑ 4 small, clear plastic cups such as 100 ml graduated medicine cups (if not evident, mark a 15 ml line on each cup)
- ❑ 5 balloons (2 red, 1 green, 1 yellow, 1 blue; each should have a diameter of 8–10 in.)
- ❑ 4 plastic straws
- ❑ 8 twist ties
- ❑ Tray
- ❑ Piece of scratch paper to use as a funnel for the baking soda
- ❑ Copy of Handout 2.1: Four Gas Samples, Observation Sheet (located on the Extras page: *www.nsta.org/climatechange*)

For each student:

- ❑ Indirectly vented chemical splash goggles
- ❑ Nonlatex aprons
- ❑ Nitrile gloves
- ❑ Copy of Handout 2.2: Four Gas Samples, Data Sheet (located on the Extras page: *www.nsta.org/climatechange*)
- ❑ Copy of Handout 2.3: Climate Change Agent Interview (located on the Extras page: *www.nsta.org/climatechange*)
- ❑ Science notebook

Safety Notes

Communicate the following safety precautions to the class on the day of the activity:

1. Personal protective equipment (indirectly vented chemical splash safety goggles, nonlatex aprons, and nitrile gloves) must be worn during the setup, hands-on, and takedown segments of the activity.

2. Students with latex allergies should not handle the balloons. If any students get dermatitis from touching latex, allow them to have a supervisory role in the lab. If it is possible that they will have a severe allergic reaction to latex particles in the air, they may need to have a video of the lab created for them while they work in another room.

3. Students should use caution when working with glassware as it can shatter if broken and cut skin.

4. Students should immediately wipe up any spilled water on the floor as it is a slip/fall hazard.

5. Students should inform the teacher immediately if they spill chemicals on themselves, someone else, the work station, or the floor.

6. Use caution when working at the rear of an automobile to collect the exhaust gases. **Anyone participating in or observing the exhaust collection should be careful not to breathe in the exhaust.** Any students who participate in or observe the exhaust collection should be under the instructor's direct supervision.

7. Students should follow the teacher's instructions for waste disposal.

8. Students should wash their hands with soap and water upon completing this activity.

Preparation

Three to Four Weeks Before You Begin the Unit

1. Obtain bromothymol blue (alternative spelling: bromthymol blue) solution from an educational science supply house or chemical lab. Bromothymol blue is available in either concentrated liquid (aqueous) or powdered form. The liquid form is easier to use. Carolina Biological carries the recommended variety. You can order it on their site (*www.carolina.com*) using the following information: Item #849167, bromthymol blue, 0.04% aqueous, laboratory grade, 4 L.

2. Collect the glass bottles. They can be obtained from a restaurant that serves bottled water in a liter bottle. You may also ask students or staff to bring in unused glass water bottles. You may want to soak the labels off. Other glass or plastic bottles may be substituted, as long as they are about as large and the neck of a balloon can fit over the top with an airtight seal. Two-liter soda bottles usually do not permit a good seal at the top, and the rough edges will pop the balloons. Test potential bottles by using one to fill a balloon with carbon dioxide as described in this activity.

Before the Day of the Activity

1. Read through the instructions for this session and practice

   a. generating carbon dioxide gas using a reaction of baking soda and vinegar, and then collecting it in a balloon;

   b. detecting carbon dioxide with BTB; and

   c. collecting car exhaust in a balloon.

2. You will need 750 ml BTB solution for each class. You can use a clear 1-liter soda bottle for each class or a 1-gallon jug to make 3.8 liters. To make a diluted solution, fill a 1-gallon (or 3.8-liter) container with 2,700 ml (nine-tenths full) of distilled or deionized water, and then add 300 ml of BTB concentrate. The solution should be a deep, transparent blue. The exact concentration is not critical. However, you should test the solution by pouring 15 ml of BTB solution into one of the small, clear 100 ml size cups. Using a straw and wearing goggles, bubble one lungful of breath through the small cup of solution. If the solution turns green, it is OK. If it stays blue or only slightly bluish-green, then it is too concentrated. Pour out some solution, add more water, and test again. If it turns yellow, it is too diluted, and you need to add more BTB.

   If you start with BTB in powdered form, the instructions are the same, except you must first prepare the concentrate. Put 1 gram of BTB powder into a 1-liter container. Add 16 ml of 0.1 molar sodium hydroxide (0.1 M NaOH), and dissolve the crystals of BTB. Add 1 liter of distilled or deionized water to make 1 liter of concentrate.

   For each group of four students, provide a squirt bottle with about 150 ml of prepared BTB solution.

3. Decide what vehicle will be used to collect the car exhaust. It must be a fully gasoline-powered vehicle, not a hybrid. It will need a round tail pipe since square ones are difficult to seal.

Note: If you have a 90-minute block of time, you can have students collect the first three gas samples (human breath, air, pure CO_2), and then take the whole class down to the car and collect the exhaust. This will require one other adult assistant to sit in the car with their foot on the brake while you collect the exhaust with students.

If you have a 60-minute block of time, you will need two adult assistants. One will remain in the classroom to supervise gas collection of the first three samples while you take the students in charge of collecting the exhaust samples and one other adult to the car.

4. Prepare a cone for collecting car exhaust by rolling up a manila folder lengthwise. One end must be larger than the tail pipe's diameter, and the other end must be small enough for a balloon to fit over it. Use plenty of tape to hold the cone in the proper shape, and make the sides of the cone airtight. Trim the ends of the cone if necessary. Make a spare cone and have masking tape, folders, and scissors on hand when you collect the gas. A demonstration video of this process is provided on the Extras page: *www.nsta.org/climatechange.*

5. Practice filling several balloons with car exhaust so you can fill them quickly and reduce the amount of time students spend near the idling car.

Safety Note: After starting the car, make sure the car is in PARK with the parking brake ON before stepping behind the car. As you and your students fill the balloon, your adult assistant will be sitting in the car with the car in PARK, the parking brake ON, and their foot on the brake, as well.

Place a balloon on the small end of the cone. Approach the exhaust pipe from the side and hold your breath while filling the balloon so you don't inhale the gases. Hold the cone over the tail pipe and push toward the car. If you have a good seal, it will take less than a second for the balloon to fill.

6. Standardize the samples by letting out a small amount of gas until the balloon just fits through the roll of masking tape. Twist the balloon and secure it closed with a twist tie. It is very important now and when you do the activity with the class to standardize the volume of gas in the balloon outside or under a fume hood in your classroom so that no one is exposed to exhaust fumes.

7. Decide which adult(s) will assist you with the exhaust collection, and arrange to have them accompany you and the class.

Note: It may not be practical for you to have a car accessible to your students at school. In this case, you can collect the gas samples at home the evening before or

the morning of the session. Be sure to overfill the balloons since leakage will occur over time.

8. Make copies of Handout 2.1: Four Gas Samples, Observation Sheet; provide one per group of four students in each class.

9. Make copies of Handout 2.2: Four Gas Samples, Data Sheet and Handout 2.3: Climate Change Agent Interview; provide one per student.

   ***Note:** You may want to provide an opportunity for differentiated instruction by having students make their own data sheets to record the results of the experiment.*

On the Day of the Activity

1. Park the car outdoors within close walking distance of your classroom. Preferably, the class should not have to cross a street to get to the car. Decide where the students should stand to watch you collect the gas samples so they can see what is happening but are not in traffic. Since auto exhaust contains toxic gases, students should stand back far enough so they do not have to breathe in car exhaust.

2. Set up **one lab tray for your demonstration** with the following materials:

   ❑ Empty 1 L glass bottle

   ❑ Squirt bottle containing the diluted BTB solution

   ❑ Teaspoon

   ❑ 100 ml graduated cylinder or measuring cup

   ❑ 8 oz. plastic cup labeled "Baking Soda," filled halfway with baking soda

   ❑ 8 oz. plastic cup labeled "Vinegar," filled with about 175 ml of vinegar

   ❑ Small, clear plastic 100 ml cup with a 15 ml line marked on it

   ❑ 3 balloons (1 yellow, 1 green, and 1 blue)

   ❑ Roll of masking tape

   ❑ 2 plastic straws

   ❑ 3 twist ties

   ❑ 2 pairs of indirectly vented chemical splash goggles (one for you and one for a student assistant)

   ❑ Nonlatex aprons

   ❑ Nitrile gloves

   ❑ Piece of scratch paper to use as a funnel for the baking soda

❏ Copy of Handout 2.1: Four Gas Samples, Observation Sheet

❏ Copy of Handout 2.2: Four Gas Samples, Data Sheet

❏ Air pump

3. Decide where to put the air pump so that it may be easily shared by each lab group as the students collect their samples after your demonstration.

4. Set up **one lab tray per four students** with the following materials:

❏ Empty 1 L glass bottle

❏ Squirt bottle containing the diluted BTB solution

❏ Teaspoon

❏ 100 ml graduated cylinder or measuring cup

❏ 8 oz. plastic cup labeled "Baking Soda," filled halfway with baking soda

❏ 8 oz. plastic cup labeled "Vinegar," filled with about 175 ml of vinegar

❏ 4 small, clear plastic 100 ml cups with 15 ml lines clearly marked

❏ 5 balloons (2 red, 1 yellow, 1 green, 1 blue)

❏ Piece of scratch paper to use as a funnel for the baking soda

❏ 4 plastic straws

❏ 10 twist ties

❏ 2 rolls of masking tape

❏ 4 indirectly vented chemical splash goggles (1 for each student)

❏ 4 nonlatex aprons (1 for each student)

❏ 4 pairs of nitrile gloves (1 pair for each student)

❏ Copy of Handout 2.1: Four Gas Samples, Observation Sheet

❏ 4 copies of Handout 2.2: Four Gas Samples, Data Sheet

Begin!

Gathering Carbon Dioxide Samples

1. Ask students why their breath is like exhaust from a car. Have them jot a few ideas in their notebooks. When they are finished, ask for a few responses.

2. Ask the class to tell you what they already know about carbon dioxide gas—what it looks like (it's colorless), how it is produced (as a by-product of human respiration, the combustion of fossil fuels and biomass, etc.), and what its properties are (it's a naturally occurring gas; at low concentrations, the gas is odorless; at higher concentrations, it has a sharp, acidic odor; it's found in carbonated drinks; etc.). Students don't need to know all of the properties; you can just use this as an assessment of their knowledge.

3. Ask students to put the following gases in order of the concentration of carbon dioxide found in each, from lowest to highest. They can write their answers in their notebooks.

    -   Human breath
    -   The air
    -   The product of a reaction of baking soda and vinegar
    -   Automobile exhaust

4. Explain that to detect carbon dioxide, the first thing you need is a good supply of gas. Ask if anyone knows an easy way to make pure carbon dioxide. ("Breathing out" might be an answer, but it doesn't produce pure carbon dioxide.)

5. Reveal that an easy way to make pure carbon dioxide is in a chemical reaction between two common substances: vinegar and baking soda. Ask students to watch the demonstration because they will be doing it themselves.

6. Write the amounts of chemicals to be used in this reaction on the board:

    -   4 heaping teaspoons (20 grams) of baking soda
    -   100 ml of vinegar

7. Ask for a student assistant to come to the demonstration area. They will pour the vinegar into the bottle. Both you and your student should wear safety goggles for the setup, hands-on, and takedown portions of the demonstration.

8. Place an empty 1-liter bottle in front of you. Demonstrate how to make a funnel from a piece of paper to pour baking soda powder into the bottle. Put 4 heaping teaspoons of baking soda into the bottle, using the paper funnel.

9. Measure 100 ml of vinegar into the graduated cylinder or measuring cup. Be ready to quickly stretch the neck of the **yellow balloon** over the top of the bottle to capture the escaping gas.

10. Have the student assistant pour the vinegar into the bottle. The reaction will push the air in the bottle out in less than a second. Quickly stretch the balloon over the top of the bottle. The balloon should inflate to about 10 cm (4 inches) in diameter.

11. Have the student help you tightly tie off the neck of the balloon with a twist tie. (Helpful hint: Twist the balloon neck immediately after collecting the gas so no gas escapes before you get a chance to tie it off.) See Figure 2.1 and Figure 2.2. Warn students not to puncture the balloon with the sharp end of the twist tie. Have your assistant return to their seat.

Figure 2.1: Twist the balloon neck immediately after collecting the gas.

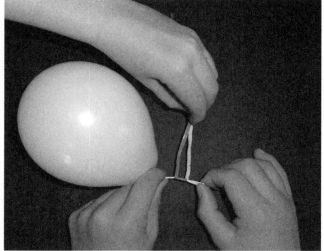

Figure 2.2: Wrap a twist tie around the balloon twice, and then twist the ends together.

12. Tell students that they will have to work as a team as they collect their own samples, with one person holding the bottle while someone else stretches the balloon over the end of the bottle.

13. Demonstrate how to collect a sample of human breath by blowing up the **green balloon** and closing it with a twist tie.

14. Demonstrate how to use the air pump to fill the **blue balloon** with air from the room to approximately the same size as the balloons with carbon dioxide and human breath. You don't need to tie off this example; just release the air from the balloon.

15. Have students number off (from 1 to 4) to determine who is "in charge" of collecting a specific gas sample. Once they have determined their numbers, assign responsibilities to students as follows:

    #1—Human breath, **green balloon** (also assistant to #3)

#2—Air (from the air pump, which will need to be shared by the class), **blue balloon**

#3—Pure carbon dioxide (with assistance from student #1), **yellow balloon**

#4—Car exhaust, **red balloons** (They will need to bring the two balloons, two twist ties, and a roll of masking tape with them to the car; they will collect two samples of gas per student in case one balloon pops or loses the gas sample.)

16. Have one student from each group pick up a materials tray. Tell students that once their tray arrives and they are wearing their safety goggles, nonlatex aprons, and nitrile gloves, they can get to work.

    Note: If you are splitting up your class in order to fit all sample collection into a 60-minute period, this would be the time to take your #4 students and your adult assistant to collect the car exhaust samples. Be sure to bring two cones (the original and your spare) plus tape, folders, and scissors. Remind #4 students to bring two red balloons, two twist ties, and a roll of masking tape.

17. Circulate around the room (or have your assistant do this if you're collecting car exhaust), helping as needed. If students lose their sample of carbon dioxide, have them add another 100 ml of vinegar to the contents of the bottle. There is enough baking soda in the bottle to generate carbon dioxide several more times.

18. **If you are *not* splitting your class,** assemble your students once the first three gas samples have been successfully collected and tied off. Be sure to bring two cones (the original and your spare) plus tape, folders, and scissors for repairs. Remind the #4 students to bring two red balloons, two twist ties, and one roll of masking tape.

Detecting Carbon Dioxide

1. While students are working, prepare the next demonstration by measuring and pouring 15 ml of liquid BTB into a small, clear plastic cup.

2. When students have finished, demonstrate how to test for the presence of carbon dioxide with the assistance of a student volunteer.

   a. Put on your safety goggles, nonlatex apron, and nitrile gloves.

   b. Start by demonstrating the test on the balloon with human breath (the **green balloon**), which is what students will test first. (If the students have difficulty with this procedure, it is easy to replace the gas sample.) Insert a straw into the neck of the green balloon and seal it tightly with a twist tie. (See Figure 2.3, p. 40.)

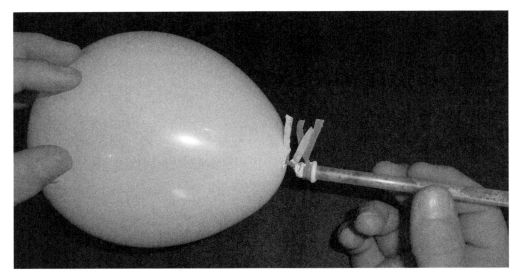

Figure 2.3: Example of how to insert a straw into the neck of the balloon. One twist tie is still sealing the balloon while the other twist tie attaches the balloon to the straw.

    c. Have your student assistant pick up the roll of masking tape. They will use this as a standardizing tool to make the gas samples uniform.

    d. Slowly loosen the twist tie sealing the balloon, releasing gas until the balloon just fits through the roll of masking tape.

    e. Clamp off the flow of gas with your fingers on the straw, and place the bottom end of the straw into the BTB.

    f. Loosening your grip, slowly bubble the gas through the blue liquid. Squeeze the balloon to release all of the gas.

    g. Observe the color of the liquid. (It should be slightly green.)

Note: Tell students that they need to take care not to get any baking soda in their cups of BTB. Since baking soda is a base, it will neutralize carbonic acid and prevent the BTB from changing to the appropriate colors.

3. Hold up Handout 2.1: Four Gas Samples, Observation Sheet. Tell students they will put this sheet on their tray and then fill five small cups with 15 ml of BTB and place them on their designated locations. Tell students they will keep the BTB solution in the cups until the end of the experiment so they can compare the colors.

4. Hold up Handout 2.2: Four Gas Samples, Data Sheet and explain how each individual is to record the results. Students will record the names of the four gas samples along the line on their data sheets, from least to most carbon dioxide.

Safety Note: Point out that while none of these chemicals are considered dangerous, it is important to practice safe lab techniques. Students will leave all BTB samples on their trays in case they spill. They should not splash any chemicals around. Remind students that they will continue to wear their safety goggles, aprons, and nitrile gloves for this activity.

5. Ask if there are any questions before starting. If not, have the students adjust the size of their balloons with the roll of masking tape and bubble the gas samples through the cups of BTB.

6. Circulate among the groups to help as needed. Watch to see that no baking soda gets into the BTB.

7. As they finish, ask students to record their results on their data sheets and write a sentence or two to describe their results at the bottom of the sheet. Once their data sheets are completed, ask them to put their equipment back on the trays and return the trays to the preparation area.

Discussing the Results

1. Ask students to compare how the samples of gas affected the BTB. Did the room air affect the BTB at all? If so, how many groups observed a color change? (The results of the experiment depend on the exact concentration of BTB and on the ventilation in the room. In most cases, the room air changes the color of BTB very little, if at all.)

2. Ask students what this means in terms of the concentration of carbon dioxide in the air. (Room air contains the minimum amount that can be detected by a BTB test.)

3. Ask, "How does BTB react in the presence of almost pure carbon dioxide?" (It turns yellow.)

4. Tell students that BTB stands for bromothymol blue solution, which can be used to test for the presence of carbon dioxide gas. It is an acid/base indicator and changes to yellow in the presence of an acid. When carbon dioxide bubbles through the solution, it creates carbonic acid, which is slightly acidic. The BTB changes color from blue to blue-green (a little carbon dioxide) to greenish-yellow (more carbon dioxide) to yellow (a lot of carbon dioxide). Draw this color scale on the board. This drawing will help students realize that green is an intermediary color between blue and yellow and that there are various other shades of color between blue and bright yellow.

5. Ask students if they were surprised by any of the results and if their initial prediction was the same as or different from their results. What can account for the differences in the relative concentrations of carbon dioxide in the air, in their breath, from a reaction of vinegar and baking

soda, and from car exhaust? They should note that there is quite a bit of carbon dioxide in car exhaust.

6. Have students reflect on the limitations of this model for demonstrating concentrations of CO_2. Observations might include the following:

    - Once the BTB has turned yellow, it can't differentiate the relative amount of CO_2 any further.

    - The experiment only allows them to measure relative amounts of CO_2, rather than specific amounts that would be possible in a titration experiment.

    - The collection methods for obtaining the gas samples aren't perfectly standardized, making room for error.

    - There may have been inconsistencies between lab teams conducting the experiment.

Climate Change Agent Interview

1. Distribute copies of Handout 2.3: Climate Change Agent Interview to each student. It features Dr. Shallin Busch, who studies ocean acidification. Dr. Busch works as a research ecologist with NOAA and has a "time machine" used to model conditions in past, present, and future oceans.

2. Allow about five minutes for students to read the interview and jot down their impressions and questions in their notebooks. After most have finished, ask, "After seeing how various carbon dioxide samples affect the pH of water, why do you think Dr. Busch is concerned about the amount of carbon dioxide in the atmosphere?" Give students time to make notes in their notebooks and then talk at their tables. Ask for comments from students as time allows.

Reviewing Statements and Questions About Climate Change

1. Have students review the lists they made in their notebooks about what they have heard about climate change. Ask if they now have enough evidence to determine the accuracy of any of these ideas. If students have enough evidence to categorize a statement as accurate, they should record it on a sentence strip and post it in the Accepted

as Accurate column. If they have enough evidence to categorize a statement as inaccurate, they should record it on a sentence strip and post it in the Accepted as Inaccurate column. If the evidence suggests that an idea on their list is accurate but not definitive, they may post it in the Needs More Information/Evidence/Research column. *In all cases, students need to support their decisions with evidence.*

Example: If students had heard that cars are a source of carbon dioxide in the atmosphere, they can write the statement on a sentence strip and post it in the Accepted as Accurate column.

Example: If they had heard that cars are *not* a source of carbon dioxide in the atmosphere, they could write this statement on a sentence strip and post it in the Accepted as Inaccurate column.

Example: If they had heard that gasoline-powered cars produce less carbon dioxide than diesel-powered semi trucks, they could write this statement on a sentence strip and post it in the Needs More Information/Evidence/Research column.

2. Ask students if they can add any new statements to the Accepted as Accurate, Accepted as Inaccurate, or Needs More Information/ Evidence/Research wall columns, based on their discussions and the evidence gathered in Session 2. *In each case, they need to support their decision with evidence.*

3. Revisit the list of Questions We Have About Climate Change. Ask students if they think they have enough information now to answer any of these questions. If so, they may be able to place the answer as a statement in the Accepted as Accurate wall column. *In each case, they need to support their decision with evidence.*

Example: Say students had asked, "How does climate change affect ocean acidification?" Students might come up with the following answer: "Carbon dioxide in the atmosphere makes ocean acidification worse." If students can support the claim with evidence, place the statement in the Accepted as Accurate column.

Example: Say students had asked a question about how much carbon dioxide comes from cars. They can initially write a statement that car exhaust contains more carbon dioxide than human breath. Challenge students to find out more specifics by using their data to calculate the actual volume of CO$_2$ produced by a car during a single trip to school or throughout an entire year. (Note that the average car is driven 10,000 miles annually.) Students can compare their results with data available online.

4. Ask students to take a minute to record anything that was surprising in today's lesson. Ask students, "Do you have any new questions that you would like to add to the Questions We Have About Climate Change column? Be ready to share your thinking with the class."

Extending the Session

1. Students may want to study the difference between traditional gasoline-powered vehicles and more "green" vehicles. One resource for this topic is the EPA's web page: *www.epa.gov/greenvehicles/greenhouse-gas-emissions-typical-passenger-vehicle.*

2. Students may want to explore the effect that plants have on CO_2 levels. Here is a simple extension to the experiment in this session that shows that even though plants take in CO_2 during photosynthesis, they give off CO_2 during cellular respiration at night.

   • Fill a test tube approximately one-third full of BTB.

   • Place a sprig of elodea into the test tube. (Use a pencil to push it all the way into the bottom of the tube.)

   • Wrap the test tube in foil so that no light can get in, simulating night, and place it in a test tube rack for at least 24 hours.

   • Unwrap the foil and note the color change. What can students conclude? Can they create an explanation for the difference in color?

Background for Teachers

The Carbon Cycle

Carbon dioxide is a colorless gas that currently makes up a little over 0.04% of the atmosphere. Another way to measure carbon dioxide is in parts per million (ppm). If we measure carbon dioxide in this way, we can determine that it recently surpassed 400 ppm. In other words, out of one million particles of atmosphere, slightly more than 400 are carbon dioxide.

Living organisms contribute to the carbon dioxide in the atmosphere through the process of cellular respiration. For every molecule of glucose that a living organism metabolizes, or burns, six molecules of carbon dioxide are released into the atmosphere. The carbon in the food we eat is part of an ongoing cycle where carbon is stored by plants through the process of photosynthesis. It is then released back into the atmosphere by cellular respiration of those plants and by other organisms as the energy stored by plants moves up through

the trophic levels. When organisms die, their bodies decay and release CO_2 back into the atmosphere, as well. The matter made from CO_2 cycles in the environment through the cellular processes of the life and death of living organisms. This is considered the cycling of *current carbon*. Burning and cutting down trees, like decomposition, returns CO_2 from living organisms back to the atmosphere, so this is also thought to be cycling *current carbon*.

This can be contrasted to adding *sequestered carbon* to the atmosphere. This occurs when carbon that has been stored for thousands or even millions of years is released into the atmosphere. One often overlooked source of sequestered carbon is volcanic eruptions. One should note that, according to the U.S. Geological Survey, the world's volcanoes, both on land and undersea, generate about 200 million tons of CO_2 annually. Other atmospheric sources of *sequestered carbon* are the melting of permafrost and the burning of fossil fuels. Human activities—mostly the burning of coal and other fossil fuels but also cement production, deforestation, and other landscape changes—emitted roughly 40 billion metric tons of carbon dioxide in 2015.

Combustion of Fossil Fuel

When we burn gasoline, we are using carbon that was sequestered by plants millions of years ago and has been converted into what we call fossil fuel. The basic formula for the reaction when we burn gasoline is: $2\ C_6H_{18} + 25\ O_2 \rightarrow 16\ CO_2 + 18\ H_2O$.

For students who have studied chemistry, you might challenge them to determine the mass of carbon released into the atmosphere for every gallon of gasoline that is burned in a car.

It will be surprising that one gallon of gas, which weighs 6.3 pounds (2,858 grams) releases 17.6 pounds (7,983 grams) of carbon dioxide or 5.31 pounds (2,404 grams) of pure carbon. This is because much of the mass of carbon dioxide comes from the oxygen that bonds with the carbon in the combustion process.

Current Carbon Versus Sequestered Carbon

Part of the debate about the amount of carbon dioxide in the atmosphere includes arguments from those who rationalize that CO_2 is "natural," because living organisms breathe it out in respiration. Of course, all carbon dioxide is natural, but the level of CO_2 in our atmosphere has been a determining factor in our climate for more than a billion years. Paleoclimatologists have concluded that CO_2 levels were five times higher when dinosaurs roamed the Earth than they are now. The seas were approximately 100 feet higher than they are now, and most of the Earth had a tropical climate. Yet, it is dangerous to imply that the existence of a CO_2 level this high in the past means it is a natural process and, therefore, not harmful. During these times, Earth had no human inhabitants, much less our current population of more than 7 billion people—many of whom live along coastlines that are particularly vulnerable to sea level rise caused by a warming planet.

For the past 800,000 years, the CO_2 level in our atmosphere has not risen above 300 parts per million (ppm) and has averaged 280 ppm. In 1850, the amount of atmospheric CO_2 was 285 ppm. In 1910, it crossed the 300 ppm mark. As of 2018, it was at 408 ppm.

During those 800,000 years, our planet has evolved the ecosystems we live in today, with a livable temperature in most places and adequate rainfall for plants to thrive. This level of CO_2 has kept the heat energy on our planet from driving the weather systems to extremes. The CO_2 in this carbon cycle is considered *current carbon,* or the amount that has existed for this 800,000-year period of balance in our ecosystems.

With the advent of the Industrial Revolution in the late 1800s, oil deposits that were *sequestered* (or buried) from 540–65 million years ago were brought up to the surface of the Earth and burned, releasing their stored CO_2 into the atmosphere. This additional CO_2 has caused the greenhouse effect to increase by more than 30%, resulting in numerous outcomes from a warming planet. Students will research these lines of evidence of global warming in Session 5.

Bromothymol Blue

Bromothymol blue (alternative spelling: bromthymol blue), or BTB, is an indicator that changes color in response to pH. At pH 7.6, BTB will be a strong blue color. As pH decreases, its color will change to green and eventually become yellow at pH 6. When carbon dioxide is added to water, the pH decreases because carbonic acid is formed according to the following formula: CO_2 (g) + H_2O (l) $<=> H_2CO_3$ (aq).

BTB provides proxy data for the amount of carbon dioxide present in the gas samples, as we are using pH to gain insight into the levels of carbon dioxide.

A limitation of this investigation is that it is not possible to quantitatively determine the concentration of CO_2. Once the BTB has turned yellow, additional CO_2 will not create further change. Pure CO_2 will elicit almost the same color as the car exhaust, which is still 71% nitrogen.

If you have access to probeware, you can either measure the CO_2 in ppm from samples blown into a biochamber or collect proxy data by measuring the pH of samples bubbled through water.

Pedagogy

The purpose of Session 2 is to build concrete experiences about sources of carbon dioxide in the atmosphere. Students compare the relative impact on CO$_2$ content in the atmosphere by measuring CO$_2$ in a specific quantity of gas gathered from a pure sample of CO$_2$, the atmosphere, human breath (representing cellular respiration of life on Earth), and exhaust from a car. You may want to have students explore diesel versus gasoline cars and other sources of CO$_2$ as determined by students. This experience is to prepare students to analyze CO$_2$ data gathered from a variety of sources and make evidence-based forecasts and use computational representations to illustrate relationships among Earth systems in future sessions.

Instead of using bromothymol blue for a qualitative measurement, you may choose to use probeware for quantitative comparisons. If you use carbon dioxide probeware and biochambers, students will need to be careful about cleaning CO$_2$ from the chamber between trials. Once you establish the relationship between carbon dioxide in water and pH, you may use pH probeware, as well. With any method, it is important to discuss the strengths and weaknesses of the method for measuring relative sources of CO$_2$ in the atmosphere.

In your discussion, deepen student thinking by using a variety of questions relating to crosscutting concepts:

- Ask students to look for patterns.
- Ask if they have enough data to establish cause and effect or correlation.
- Have students identify the significance of the scale and proportion of sources of CO$_2$.
- Can students use this model to predict the behavior of the system?

References

NASA. Global Mean CO$_2$ Mixing Ratios (ppm): Observations. *https://data.giss.nasa. gov/modelforce/ghgases/Fig1A.ext.txt* (accessed November 3, 2018).

NOAA. CO$_2$ at NOAA's Mauna Loa Observatory reaches new milestone: Tops 400 ppm. *www.esrl.noaa.gov/gmd/news/7074.html* (accessed November 3, 2018).

Scott, M., and R. Lindsey. 2016. Which emits more carbon dioxide: volcanoes or human activities? Climate.gov. *www.climate.gov/news-features/climate-qa/which-emits-more-carbon-dioxide-volcanoes-or-human-activities* (accessed November 3, 2018).

SESSION 3

The Greenhouse Effect

Introduction

The greenhouse effect is critical to understanding the basic mechanism of global warming. By adding more CO_2 (a greenhouse gas) to the atmosphere, more of the radiated heat from the Sun is trapped, warming the planet. This increase in heat adds more energy to the cycles that affect our weather, such as the water cycle and ocean-current patterns.

In this session, students are engaged by considering the danger of leaving a dog in a car on a sunny day with the windows rolled up. In order to best understand how to keep the dog safe from overheating, they first participate in a kinesthetic model of the greenhouse effect using their classmates to show how light and heat energy is affected by changes in carbon dioxide in the atmosphere.

Students then watch a video that models the greenhouse effect and evaluate the accuracy of that model. They compare their kinesthetic model to the video model and decide which model best suits their learning style for understanding the greenhouse effect. Students apply their new knowledge of the behavior of light and CO_2 in creating a system to allow the dog to remain in the car on a sunny day.

Additional data are provided to students, which includes the total CO_2 emissions from 29 countries compared with their per capita output. Students also study the CO_2 output from various economic sectors, both globally and in the United States, and consider how the data inform how we might begin to address the issue of global warming.

Objectives

1. To provide students with demonstrations and models that illustrate the greenhouse effect as a cause of global warming

2. To understand the role that carbon dioxide plays in the greenhouse effect

3. To analyze data showing the sources of greenhouse gases in the United States and the quantity produced worldwide

What You Need

Gather the following materials.

For the class:

- ❑ A method to project web pages with videos to the class; speakers are necessary
- ❑ The current list of questions and statements on the wall OR access to the electronic documents with questions and statements
- ❑ Blank sentence strips for additional questions and statements (if you're using the wall columns)
- ❑ Marking pens
- ❑ 6 sheets of colored paper (3 blue, 1 yellow, 1 red, 1 brown), at least 8.5 × 11 in.
- ❑ 3 large sheets of white paper, at least 8.5 × 14 in.

For each group of four to six students:

- ❑ Laptop or electronic device with internet access

For each student:

- ❑ Copy of Handout 3.1: Gas Composition of the Earth's Atmosphere
- ❑ Copy of Handout 3.2: International CO_2 Levels in 2015
- ❑ Copy of Handout 3.3: U.S. and Global Greenhouse Gas Emissions by Sector
- ❑ Copy of Handout 3.4: Climate Change Agent Interview
- ❑ Science notebook

  Note: All handouts are found on the Extras page: www.nsta.org/climatechange.

Preparation

Before the Class

1. Using a marking pen or photocopier and 8.5 × 11 in. sheets of colored paper, make signs that can be read from the back of the class:

Yellow	**Visible-Light Photon**
Red	**Infrared Photon**
Blue	**Carbon Dioxide Molecule** (three copies)
Brown	**Rock Molecule**

2. Using a marking pen or photocopier and 8.5 × 14 in. or larger white pieces of paper, make three signs that say the following:

    Outer Space

    The Atmosphere

    The Earth

3. Decide on a large open space for the demonstration. You will need a space about 20 feet long by 15 feet wide. You can rearrange desks in your classroom or use a larger space indoors or outdoors. Post the following signs so they occupy one-third of the space equally (if you are outdoors with no posting abilities, you can select three students to hold the signs at the edge of the demonstration space):

    Outer Space   **The Atmosphere**   **The Earth**

4. Review the greenhouse effect computer model from Stile Education (*www.youtube.com/watch?v=fYqdKiT0Eqo*) to expand your students' understanding of the greenhouse effect after they have experienced the physical demonstration.

Begin!

Introducing the Greenhouse Effect Demonstration

1. Ask students why it is dangerous to leave a dog in a car with the windows rolled up on a sunny day. Have them record their thinking in their notebooks. They may use words, pictures, or diagrams to help explain their answers. Have them include what they have heard about light energy.

2. When students have finished, tell them they will now gain some information that may help them revise their thinking—and keep the dog safe!

3. Ask students if any of them have heard about the greenhouse effect. They may refer to their notes from Session 1. Ask for a few students to share what they have heard. Refrain from correcting any misconceptions at this time.

4. Tell students they will create a model of the greenhouse effect with their classmates serving as molecules and photons. Remind them that all models are inaccurate to some degree.

5. Explain that scientists have discovered that the energy from the Sun is composed of tiny packets of energy called *photons*. There are different sorts of photons, but two are especially important to know about for the demonstration:

   a. Photons we can <u>see</u> are called *visible-light photons*.

   b. Photons we can <u>feel as warmth</u> are called *infrared photons*.

6. Briefly review the concept that all matter is made up of *molecules,* and molecules are made of even smaller bits called *atoms*. In gases and liquids, molecules are free to move around; in solids, they are held together more rigidly.

7. Ask a volunteer to share what kinds of molecules make up the air. Pass out a copy of Handout 3.1: Gas Composition of the Earth's Atmosphere. After students have had time to look over the data, ask if anything surprises them. Many will be surprised that CO_2 only comprises 0.04% of atmosphere or that nitrogen comprises most of the atmosphere.

The Interaction Between Photons and Molecules

1. To begin the demonstration, point out the signs that divide the open area of the room or outdoor space into three regions: Outer Space, The Atmosphere, and The Earth.

2. Ask for three volunteers.

   a. Ask one student to stand in the **Outer Space** area, and hand them a yellow sign labeled "Visible-Light Photon."

   b. Ask a second student to stand in **The Atmosphere** area, and hand them a blue sign labeled "Carbon Dioxide Molecule."

   c. Ask a third student to stand in **The Earth** area, and hand them a brown sign labeled "Rock Molecule." Also provide a sign labeled "Infrared Photon," but have the student hold that sign behind the Rock Molecule sign.

   d. Ask all three volunteers to hold up their signs so the whole class can read them. See Figure 3.1 for an example.

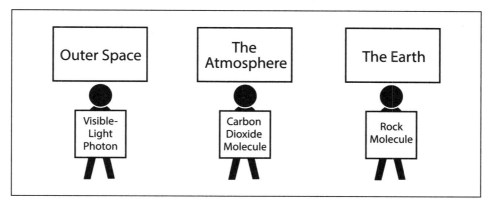

Figure 3.1: Initial setup for modeling the greenhouse effect

3. Ask the student acting as the visible-light photon to slowly walk in a straight line past the carbon dioxide molecule in the atmosphere and approach the Earth.

4. As the visible-light photon does this, tell the class that the visible-light photon originated in the Sun, traveled in a straight line to the Earth, and is now entering and passing through the Earth's atmosphere.

5. Since carbon dioxide **does not** absorb visible light, the visible-light photon passes right by the carbon dioxide molecule and approaches the Earth.

6. Ask the student acting as the visible-light photon to stop at the Earth. Explain to the class that two things can happen when a visible-light photon encounters Earth's surface. Either it will be reflected back into space or absorbed by Earth and then radiated back into the atmosphere.

7. Let's suppose the energy from the photon is absorbed. Tell the student acting as the visible-light photon to hand the Visible-Light Photon sign to the student acting as the rock molecule. The rock molecule will absorb this energy, causing it to vibrate at a higher rate. Ask the student acting as the rock molecule to demonstrate this by jiggling the Rock Molecule sign. (You may want to note that atoms and molecules never stop vibrating, even in solids—but this is not illustrated in this model, which is one of its limitations. However, all models are inaccurate to some small or large degree.)

8. Mention that you cannot see molecules moving in warm substances because the molecules are too small to see. However, if you touch the rock, your skin senses this vibration as warmth.

9. Explain that, after a while, the rock molecule will cool off, or reduce vibrating significantly. When it does, it emits infrared radiation, or heat.

For example, if a rock has been in the sunlight all day, we can put our hands a few inches away from it and feel its warmth. What we feel are billions of infrared photons being given off by the rock molecules as the rate of vibration decreases.

10. To demonstrate this, ask the rock molecule to stop vibrating (jiggling the sign) and hand the Infrared Photon sign to the student who was holding the Visible-Light Photon sign (and who should still be standing nearby). Now acting as the infrared photon, this student will move away from the Earth to show the photon leaving.

11. Ask the student acting as the infrared photon to stop next to the carbon dioxide molecule on the trip away from the Earth.

12. Explain that carbon dioxide molecules **absorb** infrared photons easily. Ask, "What should our two actors do when they come near each other?" *(Answer: The student playing the infrared photon should hand the Infrared Photon sign to the student playing the carbon dioxide molecule, who will jiggle the Carbon Dioxide Molecule sign to show a higher energy level.)*

    Ask, "What should our actors do when the carbon dioxide molecule cools off?" *(Answer: The student playing the carbon dioxide molecule should stop jiggling the sign and hand the Infrared Photon sign back to the student playing the infrared photon.)*

13. Explain to the class that, at this point, the infrared photon might head off in any direction: out into space; sideways; or back to the ground, where it might be absorbed by molecules of another rock or another carbon dioxide molecule. See Figure 3.2.

14. Ask students to predict what they think will happen if more carbon dioxide is added to the atmosphere. Use the think/pair/share technique. Students can think first and then tell their neighbors their predictions. Ask for a few predictions before continuing with the demonstration.

15. Ask for two more volunteers to become carbon dioxide molecules. Hand signs to them and place them in The Atmosphere section, standing in a row near one another. Say that you are going to demonstrate what happens when additional carbon dioxide is added to the atmosphere.

16. Restart the demonstration by having the student playing the visible-light photon return to the Outer Space section. Tell the class that another visible-light photon has been emitted by the Sun, and it is now traveling in a straight line to Earth. The photon heads toward the atmosphere, where it will encounter a carbon dioxide molecule again and go right past it. Remind the class that *visible light* can pass right by a carbon dioxide molecule without warming it.

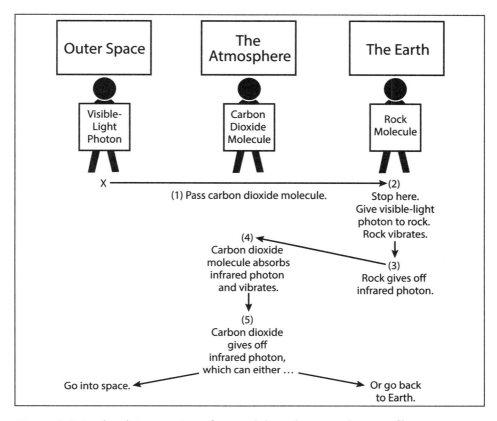

Figure 3.2: In-depth instructions for modeling the greenhouse effect

17. Have the visible-light photon reach the rock molecule again. This time, the photon is absorbed (thus the Visible-Light Photon sign should be handed to the student playing the rock molecule), causing the rock molecule to vibrate at a higher energy level. An infrared photon is emitted and goes out into the atmosphere (so the rock molecule actor gives the Infrared Photon sign to the student playing the photon). The infrared photon is absorbed by a carbon dioxide molecule, causing the molecule to vibrate at a higher energy level (the carbon dioxide molecule actor shakes the Carbon Dioxide Molecule sign). As the carbon dioxide molecule slows its vibration, an infrared photon gets radiated back to Earth, where it hits the rock molecule again, causing it to vibrate at a higher energy level and return heat into the air. The infrared photon is then released back into the atmosphere as radiant heat.

18. Once the infrared photon is emitted back into the atmosphere, it will most likely be absorbed by another carbon dioxide molecule, since there are so many in the atmosphere at this time. The carbon dioxide

molecule will vibrate at a higher energy level and then emit an infrared photon, which is radiated back to Earth.

19. Thank the volunteers and have the class take their seats.

20. At their tables, have students come up with an explanation of what happens when additional carbon dioxide is added to the atmosphere. What do they think will happen to the temperature on Earth? They should write their answers in their notebooks. After a few minutes, use the think/pair/share technique to have students share their ideas. Ask for a few volunteers to share their explanations with the whole class.

    Note: Students should understand that the chance of an infrared photon reflecting back into outer space is reduced when you add more carbon dioxide to the atmosphere. As each infrared photon encounters a carbon dioxide molecule in the atmosphere, the carbon dioxide molecule warms up the atmosphere and reflects the photon back to Earth, where it also warms the Earth.

21. Have students talk at their tables about what happens when photons of energy from the Sun interact with matter. Have them add this to their notebooks. Their understanding of this concept will be influenced by their prior knowledge and grade level.

Evaluating Additional Models of the Greenhouse Effect

1. Show "The Greenhouse Effect," a computer-generated model of the greenhouse effect, to your students (*www.youtube.com/ watch?v=fYqdKiT0Eqo*). After watching it as a class, have students watch it again on laptops or other electronic devices at their tables in small groups. Have them evaluate the model and describe where it is accurate and where it is inaccurate. Have them connect the students who played roles in the demonstrations to figures in the video.

2. After students have completed their evaluation at their tables, ask for them to share some of their findings.

3. Show the video to the entire class again. Ask for comments on where the model is accurate and where it is inaccurate. Stop the video where necessary to allow students to explain their thinking.

4. Ask students which model was most useful for them to understand the greenhouse effect—the video simulation or the physical simulation. Why? Give them a few minutes to reflect and answer the questions in their notebooks. Then ask, "What elements of your preferred model made it more effective for you?" Ask for students who chose different models to share their responses.

5. Now revisit the dog that is sitting in the car in the sunlight. Have students add their new understanding of light to their initial response.

Ask, "How is the car like a greenhouse?" Have them compare the car with its windows up to the Earth and its atmosphere. Help students understand that the glass works like carbon dioxide molecules—it lets visible light in, but it can also trap infrared photons radiated by the inside of the car, causing the vehicle to heat up. Have students create a way for the dog to be comfortable in the car on a sunny day by applying their new knowledge of light photons, infrared photons, absorption, reflection, and the transmission of photons. Have students use diagrams and descriptions to explain their thinking.

Note: This may be a good place to end the session and continue the next day.

Checking for Understanding

1. Ask, "How does carbon dioxide in the atmosphere warm the Earth?" Have students write their answers in their notebooks. They can use diagrams and words to explain. *(Answer: Carbon dioxide molecules absorb infrared photons, thus warming the atmosphere. When these molecules cool, they give off infrared photons, some of which go back to the Earth's surface. Thus, some photons are absorbed several times before escaping into space.)*

2. If students have difficulty making such generalizations, ask a series of focused questions, such as the following:

   a. When a visible-light photon is absorbed by a molecule, what happens to the molecule? *(Answer: It vibrates at a higher rate.)*

   b. What happens when that molecule cools off? *(Answer: It vibrates at a lower rate and gives off an infrared photon.)*

   c. What happens to the infrared photon? *(Answer: It radiates in a random direction: sometimes it goes into space; sometimes it goes into the atmosphere; sometimes it heads back to the Earth.)*

   d. How many times can an infrared photon warm the Earth when there is carbon dioxide in the atmosphere? *(Answer: unlimited times)*

3. Ask students to imagine that the Earth is the size of their heads. Ask, "In this model, how thick would the atmosphere be?" Have students make a prediction by spreading their hands and arms as wide as they think the atmosphere would be around a head-size Earth. Start with the largest prediction, which could be up to the full width of their spread arms, and match that prediction with your own arms/hands. Tell them you will now reveal the thickness of the Earth's atmosphere, if it were as big as a human head. Slowly move your hands together, and have students put their hands down when you have passed their prediction. Once you have put your hands together, reveal that, in this model, the thickness of the Earth's atmosphere is the thickness

of one human hair! Tell students that this should help them better understand the limited amount of atmosphere we have surrounding our Earth. Remind them that the height of the atmosphere in the video was inaccurate but necessary for a visual representation of the greenhouse gases.

Greenhouse Gases—From Global to Local

1. Hand out the two data sheets, Handout 3.2: International CO_2 Levels in 2015 and Handout 3.3: U.S. and Global Greenhouse Gas Emissions by Sector. Give students a few minutes to look over the data. They can take notes in their notebooks about the information found in the graphs. Those who finish early can look over their notes on what they have heard and determine what statements have enough evidence to be moved into any of the columns in your ongoing list of questions, statements, and further information needed.

2. When students are ready, hold a discussion about the new information. Ask for any initial impressions of what the data tell them. Clarify that international shipping accounts for the amount of CO_2 emitted annually by the transport of goods by ship. International aviation accounts for all of the CO_2 emitted by airplanes each year.

3. Draw their attention to the graph on Handout 3.2: International CO_2 Levels in 2015. What does this information tell them? Does anything surprise them? *(Possible answer: They may comment that Australia puts out a relatively small amount of CO_2 as a country but emits the most per capita of any country, even the United States. This is due to the large amount of coal used for energy production in Australia.)*

4. Now have students focus on the amount of CO_2 emitted per country. What does this tell them? *(Possible answer: They may conclude that China and the United States are the top two emitters of CO_2 by far.)*

5. If your students don't mention it on their own, ask, "What will happen to the total amount of emissions from India and China when their populations increase dramatically, as they are predicted to do?"

6. Ask students to think about the amount of CO_2 put into the atmosphere by shipping and aviation. How does this figure into our study of greenhouse gas emissions?

7. Now draw their attention to Handout 3.3: U.S. and Global Greenhouse Gas Emissions by Sector. Give students a few minutes to understand the data. Ask what the information in the graphs tells them about the source of the largest amount of greenhouse gases produced in the world. How about the second largest? Third largest? What does the graph tell them about the source of the least amount of greenhouse

gases? How does this compare with the sources of greenhouse gases in the United States?

8. Have students turn to a partner and discuss what changes we might make as a country if we are going to reduce our greenhouse gas emissions. What should we do on an international level? After a few minutes of discussion, regain the attention of the class and accept a few responses. Keep the discussion short at this time. If students start to get into a deeper discussion, ask them to record their questions or comments in their notebooks and use them to drive the research they will be conducting in Session 5.

Climate Change Agent Interview

1. Pass out Handout 3.4: Climate Change Agent Interview. It features Dr. Heidi Roop, a research scientist and strategic communications lead with the University of Washington Climate Impacts Group. Give students a few minutes to read the interview, and then have them jot their reflections in their notebooks.

2. Ask students for any impressions they had of this interview. What questions would they ask Dr. Roop if they could meet her in person? What was the most important thing they learned from her interview?

3. Ask for any other comments before moving on to the statement review.

Reviewing Statements and Questions About Climate Change

1. Ask students to review the list they made in their notebooks about what they had heard about climate change. Ask if they now have enough evidence to determine the accuracy of any more of these ideas. If students have enough evidence to categorize a statement as accurate, they should record it on a sentence strip and post it in the Accepted as Accurate column. If they have enough evidence to categorize a statement as inaccurate, they should record it on a sentence strip and post it in the Accepted as Inaccurate column. If the evidence suggests that an idea on their list is accurate but not definitive, they may post it in the Needs More Information/Evidence/Research column. *In each case, students need to support their decision with evidence.*

2. Revisit the statements posted during the previous session. Ask if anyone thinks a statement should be moved to a different column

based on new information they learned during this session. Students may decide that they need more evidence for a statement that they had previously placed in the Accepted as Accurate category, or they may now have enough evidence to move a statement that had been placed in the Needs More Information/Evidence/Research column to either the Accepted as Accurate or the Accepted as Inaccurate column. *In each case, they need to support their decision with evidence.*

3. Ask students if they can write new statements for any of the columns, based on new knowledge gained from this session. *In each case, they need to support their decision with evidence.*

4. Ask students to review the questions that are posted. Are there any questions that they can now answer with their new knowledge? If so, remove the question, write out the answer on a sentence strip, and post this statement in the Accepted as Accurate column. *In each case, they need to support their decision with evidence.*

5. Finally, ask students if they have any new questions to add to the Questions We Have About Climate Change column. Provide a few minutes for discussion and for posting questions.

Extending the Session

1. If you like to use active games as demonstrations, you may want to choose a game called The One Degree Factor to demonstrate the greenhouse effect. This interactive model incorporates both constructive and destructive activities humans do that affect the amount of greenhouse gases in our atmosphere. You can find the game on the PBS website: *www.pbs.org/strangedays/educators/season1/ag_odf_cogame.html.*

2. To help students understand how increasing the already scant amount of CO_2 in our atmosphere can make a big difference, watch an engaging animated video titled "Steroids, Baseball, and Climate Change" (*www.youtube.com/watch?v=MW3b8jSX7ec*). The video compares baseball players taking steroids to increases in CO_2 in our atmosphere. It demonstrates that there is a small amount of CO_2 occurring naturally in the atmosphere just as there is a small amount of steroids occurring naturally in our bodies. Even a slight increase in each of these substances can make a big difference.

3. Show the amusing video "Global Warming or None Like it Hot," which is used to illustrate the greenhouse effect in Al Gore's film *An Inconvenient Truth* (*www.youtube.com/watch?v=OqVyRa1iuMc*). After

viewing the video, have students determine what is accurate and what is inaccurate about the cartoon model.

4. Use this short, simple video called "How Does CO$_2$ Trap Heat?" to demonstrate the role of greenhouse gases in creating the greenhouse effect: *https://ourclimateourfuture.org/video/how-does-co2-trap-heat*. It may be helpful for younger students or those with difficulty understanding the more complex models.

5. Students can be challenged to create their own physical models of the greenhouse effect, or use this example: *www.esrl.noaa.gov/gmd/outreach/ lesson_plans/Modeling%20the%20Greenhouse%20Effect.pdf*.

Background for Teachers

The Greenhouse Effect

The initial scenario about the dog in the car is designed to uncover students' thinking and assess their current understanding of transfers of light energy. Students may have several ways to present their understanding but should be able to show that visible light can be transmitted through the glass, absorbed by the dark surfaces of the car's interior, and then radiated by the interior surfaces of the car as infrared. The infrared light is not transmitted back out through the glass but is instead reflected back into the car. In this way, infrared energy is trapped in the interior of the car, and this is why the car interior warms.

In the simulation, students will model several things. First, they will model that molecules in the atmosphere are transparent to visible-light photons, meaning the photons can pass through without being absorbed. When the visible-light photons strike the Earth, they are either absorbed or reflected. If the visible-light photons are absorbed, students can model the energy transfer of the photon to the rock molecules by vibrating their paper. When the energy of specific wavelengths of a visible-light photon is absorbed (those wavelengths that are not absorbed will be reflected as the color of the object), they activate individual electrons in individual rock molecules. The light energy moves an electron in a molecule to a higher energy level. The molecule is now said to be in an excited state. The electron will soon drop back to its original energy level, or ground state, and emit the light energy as an infrared photon. This is what makes the rock feel warm. This may be called radiant heat.

The infrared photons are emitted in all angles. Many will head back out into the atmosphere. The gases that we call greenhouse gases are not transparent to infrared photons. If an infrared photon strikes a molecule such as carbon dioxide, it will be absorbed. This will cause the molecule to move from the ground state to the excited state. Again, we model this with vibration. As the molecule drops back to ground state, it emits the infrared photon in a random direction. This infrared photon may be absorbed and emitted by other greenhouse gases or by the Earth many times before it eventually radiates back into space. Essentially, greenhouse gases trap the light energy as heat in the atmosphere.

An actual greenhouse acts in much the same way. The glass is like greenhouse gases: Transparent to light photons, it lets light energy enter the greenhouse. The light energy is absorbed by the surfaces in the greenhouse, similar to the way light energy is absorbed in the Earth and atmosphere model. Also similar to the Earth and atmosphere model, the energy is released by the surfaces in the greenhouse and emitted as infrared photons. Much like carbon dioxide, the glass does not transmit the infrared photons, so they are trapped in the greenhouse, which warms up.

As your students compare the simulation to other models of the greenhouse effect, they may notice that carbon dioxide molecules are the only greenhouse gas represented in the model they acted out. Other greenhouse gases include water vapor, methane, and other gas molecules that have chemical bond configurations that absorb the infrared photons. Students may also notice that the model does not include reflective surfaces on the Earth like ice and snow. As some students research the albedo effect in Session 5, they will learn the importance of these reflective surfaces in regulating the temperature of the Earth.

Greenhouse gases are important. If it were not for the small amount of carbon dioxide and other greenhouse gases in the atmosphere, the Earth would be in a continuous ice age. Carbon dioxide helps maintain the Earth's temperature at a comfortable level. Life on Earth has always depended on the greenhouse effect.

Pedagogy

This section focuses on the science and engineering practice of modeling. Students are introduced to a model where they take the roles of elements of the global heat regulation system, and they are asked to use the model to illustrate and predict outcomes as they manipulate individual components within the system. You may extend the use of models by using online simulations and having students evaluate the merits and limitations of the various models. In their discourse, students should be able to discuss relationships among variables and between systems and their components in the natural world.

In your discourse with students, use crosscutting concepts to deepen their thinking. There are a number of ways that you can do this:

1. As students move from the interactive models looking at data, you may challenge them to identify whether the relationship between greenhouse gases and global warming is a correlation of cause and effect. They will use this information for later consideration when they design changes to systems to cause a desired effect.

2. Students should be able to recognize that models can be used to predict the behavior of a system, but guide them to consider that predictions are limited by the precision and reliability of the assumptions used in the model.

3. Students are asked to model energy flow into and out of the system. This is an opportunity for students to connect to prior knowledge of the cycling of matter or to set the stage for further investigations of the cycling of matter. This is also a good place to begin building understanding of the effect of positive and negative feedback on a system.

Assessment Opportunities

If you want to assess individual students' understanding of the greenhouse effect, you may choose to use The Greenhouse Effect formative assessment probe from *Uncovering Student Ideas in Earth and Environmental Science* (Keeley and Tucker 2016). The probe can be found on the Extras page: *www.nsta.org/ climatechange.*

Resource

Humboldt State University has an excellent page on the difference between transmission and absorption of light: *http://gsp.humboldt.edu/olm_2016/Courses/ GSP_216_Online/lesson2-1/atmosphere.html.*

References

Keeley, P., and L. Tucker. 2016. The greenhouse effect. In *Uncovering Student Ideas in Earth and Environmental Science*, 165–169. Arlington, VA: NSTA Press.

Khan Academy. Spectroscopy: Interaction of light and matter. *www.khanacademy. org/science/chemistry/electronic-structure-of-atoms/bohr-model-hydrogen/a/spectroscopy- interaction-of-light-and-matter* (accessed on October 30, 2018).

Stile Education. 2017. "The Greenhouse Effect." YouTube video. *www.youtube.com/ watch?v=fYqdKiT0Eqo* (accessed November 3, 2018).

Strumer, J., T. Leslie, M. Liddy, and C. Gourlay. 2015. What the world's 15 biggest emitters are promising on climate change. ABC News Online. *www.abc.net.au/ news/2015-08-11/climate-change-what-top-15-emitters-are-promising/6686548.*

SESSION 4

Fact or Phony? Scientifically Evaluating Data

Introduction

In a world where fictional information is often promoted as fact and cherry-picked data are offered as evidence, it can be very difficult for students to understand how to decide if information is accurate, verifiable, from a reputable source, and complete.

This session begins by looking at a purely fictional web page cleverly disguised as factual. Students are challenged to learn important information from this web page to help them decide if the Pacific Northwest tree octopus is, in fact, in danger of extinction due to the effects of climate change. In reality, this species doesn't exist. It was invented as an internet hoax in the late 1990s. After a bit of time exploring the web page, the teacher reveals that both the species and its plight are fictional and asks for student responses when they find they have been fooled.

Students are then asked to review two sets of data and evaluate their validity as well as interpret the information they are trying to convey. Through small-group discussions, students share their perceptions of the data sets and what they mean. They are challenged to decide if this information is valid and helpful in building an understanding of global warming and climate change.

The next step is for students to be given a series of common misconceptions about climate change and global warming. Using reputable sources of data, they

dispel the misconceptions with accurate information and decide how people could be swayed into believing the misconception.

Objectives

1. To encourage students to think critically about data and carefully evaluate its relevance

2. To compare multiple sources of data, looking for concepts supported by verifiable evidence

3. To teach students how to participate in thoughtful, scientific discourse, backing up their statements with evidence that can be verified

What You Need

Gather the following materials.

For the class:

- ❑ Technology to project a web page for the class to view

- ❑ The current list of questions and statements on the wall OR access to the electronic documents with questions and statements

- ❑ Sentence strips for additional statements (if you're using the wall columns)

- ❑ Marking pens (if you're using the wall columns)

For each group of students:

- ❑ Laptop or electronic device with internet access

For each student:

- ❑ Science notebook

- ❑ Copy of Handout 4.1: How to Determine if Information Is Accurate

- ❑ Copy of Handout 4.2: Atmospheric CO_2 Levels in the Recent and Distant Past

- ❑ Copy of Handout 4.3: Looking Critically at Data

  Note: Handouts 4.2 and 4.3 can be copied back-to-back.

- ❑ Copy of Handout 4.4: Scientific Discourse Circle—Is Climate Change Caused by Human Activity?

- ❑ Copy of Handout 4.5: Lines of Evidence (3 pages)

❑ Copy of Handout 4.6: Explanations for Commonly Held Misconceptions About Climate Change and Global Warming

❑ Copy of Handout 4.7: Climate Change Agent Interview

Note: All handouts are found on the Extras page: www.nsta.org/climatechange.

Preparation

Before the Class

1. Make copies of the handouts, one for each student.

2. Preview the web page on the Pacific Northwest tree octopus (*https://zapatopi.net/treeoctopus*) as an example of a completely fictional web page designed to look factual. Decide how you will project the web page for students to view.

3. Set up a way to project or post these six commonly held misconceptions used in the Explaining Misconceptions About Climate Change section (pp. 72–73), so the whole class can see them:

   • It's been warmer before.

   • Increases in solar activity are causing the Earth to warm.

   • The Arctic is gaining ice.

   • It hasn't warmed since 1998.

   • Increasing CO_2 has little or no effect.

   • Scientists don't agree that humans are causing the Earth to get warmer.

Begin!

Can We Trust What We Read on the Internet?

1. Tell students that you want to show them an example of an animal that may be significantly affected by climate change. Direct their attention to the web page you have projected on the screen in front of the classroom, or have them look up the web page on their laptops or electronic devices.

2. Give students a few minutes to explore the web page. Ask students to find evidence that the tree octopus could be negatively affected by climate change. Have them explain their reasoning.

Note: In 2006, the University of Connecticut Neag School of Education asked 25 seventh graders from middle schools across the state to review a website devoted to a fictitious endangered species, the Pacific Northwest tree octopus. The results troubled them:

- All 25 students fell for the internet hoax.

- All but one of the 25 students rated the site as "very credible."

- Most struggled when asked to produce proof—or even clues—that the website was false, even after the UConn researchers told them it was.

- Some of the students still insisted vehemently that the Pacific Northwest tree octopus really exists.

The students, identified as their schools' most proficient online readers, were taking part in a federal research project funded by a $1.8 million grant from the U.S. Department of Education. For more information on this study, visit the web page: http://advance.uconn.edu/2006/061113/06111308.htm.

3. Ask students to determine if the web page is from a reputable source and, if so, how they determined its validity. Give them a few minutes to come up with answers. At this point, some students may notice oddities such as Sasquatch being listed as a predator to the tree octopus under the Why It's Endangered section on the About tab. The images and descriptions on the Sightings tab can be quite comical but potentially believable by students. The web page owner claims to be affiliated with the Kelvinic University branch of the Wild Haggis Conservation Society. That may or may not trigger a skeptical response.

4. Reveal to your students that this web page was created to show how easy it is to put completely false information on the internet and make it look believable. If students still maintain that the web page is true (as many of the students in the UConn study did), ask some volunteers to point out information that could be incorrect. Some quick internet searching may be necessary to sort out fact from fiction. You may want to direct students to fact-checking websites. (Snopes.com, for instance, has an interesting analysis of this false web page.)

5. Distribute Handout 4.1: How to Determine if Information Is Accurate, one for each student. Go over each item on the list to clarify its meaning. Then have students revisit the Pacific Northwest tree octopus web page and see how many examples they can find from the checklist. Depending on how many devices you have for students, they can work as individuals, in pairs, or in table groups.

6. After about 10 minutes, reconvene the class and ask students for examples of what they found. After a number of inaccuracies have been revealed, ask students what they have learned about checking for

accurate information on the internet. Tell them they will use this list to evaluate information they research in this unit.

Effective Data Analysis—CO$_2$ Levels From Our Recent and Distant Past

1. Explain that students will now evaluate two sets of data showing CO$_2$ measurements taken in two ways. Pass out copies of Handout 4.2: Atmospheric CO$_2$ Levels in the Recent and Distant Past and Handout 4.3: Looking Critically at Data (one page if copied back-to-back). Give students a few minutes to review the graphs on Handout 4.2. Then ask, "Which graph do you think has a more reliable source? Why?" In Graph 1, data were gathered by direct observation; in Graph 2, data were gathered by "proxy data collection." The direct-observation data was obtained at the Mauna Loa Observatory on the big island of Hawaii with instruments that directly measure the CO$_2$ content of the atmosphere. Dr. Charles David Keeling began taking observations in 1958, and conditions are monitored to this day. The proxy data was taken from two different ice cores, one in Greenland and one in Antarctica. The data from the oldest records came from a core that was drilled 2,803 meters (6,834 feet) into the Antarctic ice shelf. Bubbles of gas were trapped in the ice, allowing scientists to sample ancient air and determine the amount of CO$_2$ in the atmosphere over 400,000 years ago! Scientists correlate the CO$_2$ from ice core samples with other proxy data sources to ensure reliability.

2. Now direct students' attention to Graph 1 and ask a volunteer to identify what the scale on each axis represents. *(Answer: On the horizontal axis, the scale goes from the year 2005 to 2018; on the vertical axis, the scale represents the amount of CO$_2$ in parts per million.)*

3. Ask students to take a few minutes to look over the Graph 1 data and record their thinking on Handout 4.3: Looking Critically at Data. Students should note what the data tell them, what the data *does not* tell them, and what else they want to know to better understand what the graph reveals.

4. Next, have students look at Graph 2 and explain what the scale on each axis represents. Make sure that students recognize that the scale on the vertical axis of Graph 1 is different from the scale on Graph 2.

*(Answer: On the horizontal axis, the scale goes from 400,000 years ago to the year 1950, represented by 0. The present [about 70 years after 1950] is represented just on the other side of 0. This is a small amount on a scale of 0 to 400,000. [**Note:** Since this graph goes in reverse of traditional measurements, with 0 being on the far right of the axis rather than at the beginning, make sure students understand this difference.] For the vertical axis, the scale represents the*

amount of CO_2 in parts per million, recorded in December 2017 as 407.62 parts per million, or 0.040762% of our atmosphere.)

5. Ask students to take a few minutes to look over the Graph 2 data and record their thinking in Handout 4.3 about what the data tell them, what the data *do not* tell them, and what else they want to know to better understand what the graph reveals.

6. If students do not know when we started burning fossil fuels (coal, oil, natural gas, etc.) to power vehicles and machinery, mention that it was during the Industrial Revolution, a period from 1750 to 1850. During this era, changes in agriculture, manufacturing, mining, transportation, and technology had a profound effect on the social, economic, and cultural conditions of the times. It began in the United Kingdom and then spread throughout Western Europe, North America, Japan, and eventually to the rest of the world.

Scientific Discourse Circle

1. Divide the class into teams of about four students each to begin the Scientific Discourse Circle[1]. Pass out Handout 4.4: Scientific Discourse Circle—Is Climate Change Caused by Human Activity?

2. Go over the sheet with the class and have each student jot down their own ideas about the statements in the space provided on the data sheet.

3. Once students have finished, tell them they are going to discuss the statements with their team. Each member should participate and contribute to the discussion. Like real scientists debating an issue, they may not agree on all points. Emphasize that this is OK!

4. Explain the procedure for the Scientific Discourse Circle:

    a. For each statement, one student states what they think and *provides specific evidence* to support this reasoning. No other members of the group may add their comments when a student is speaking. This is a listening activity first—not a group discussion.

    Note: This part may be difficult for some students who are not used to supporting their statements with evidence.

    b. Each student takes a turn either agreeing or disagreeing with the statement (*not* with their fellow students) and supporting this reasoning with evidence.

---

[1]The Scientific Discourse Circle is inspired by similar group activities for students in the Seeds of Science/Roots of Reading curriculum, copyrighted by The Regents of the University of California and used here with permission.

c. After each student has had a chance to speak, the group tries to come to an agreement on the correct answer. All discourse needs to be respectful. It is not necessary for all students to agree. In fact, this is often how science moves forward, with one dissenting voice causing the scientific community to look again at the evidence and perhaps draw new conclusions. Good scientists are open to changing their minds based on evidence.

Note: The main point of the Scientific Discourse Circle is for students to think about and discuss ideas and evidence in order to find the best explanation. Scientists are able to listen to the reasoning of others and change their minds if they think their viewpoint is no longer supported by the evidence or if another explanation is better supported by the evidence. At the same time, scientists are not easily swayed by arguments that are not based on evidence. They must decide for themselves if a particular bit of evidence supports or does not support a particular explanation or position.

5. Let students know that scientists rely on evidence. They try to answer the question "What explanation best matches all of the available evidence?"

6. Once the groups have had a chance to discuss their views—first individually, then in a group—ask for volunteers to share how the process went for their groups. Some sample questions:

   • Did any group have all participants agree from the beginning?

   • Did any students change their minds? What evidence caused them to change their minds?

   • Did any groups remain divided? If so, what was the issue that kept them divided?

7. Some students may decide that there is not enough evidence in Graph 1 to draw scientific conclusions. Applaud their reasoning! Graph 1 is specifically included to provide students with short-term data that may not be sufficient for them to be able to fully answer the questions on the handout. This graph is designed to demonstrate what cherry-picked data look like, and it can be used as an example during the next activity in this session. Remind students that they learned in Session 1 that scientists agree the minimum number of years to collect data for evaluation of climate is 30 years. If that was not clear, share this definition:

   Climate: *Climate* is usually defined as the "average weather," based on collecting statistics over a period of time ranging from months to thousands or millions of years. The classical minimum period is 30 years, as defined by the World Meteorological Organization (WMO). These statistics are most often surface variables, such as

temperature, precipitation, and wind. A simple way of remembering the difference is that climate is what you expect (e.g., cold winters) and weather is what you get (e.g., a blizzard).

Explaining Misconceptions About Climate Change

1. Tell students that they are going to look over some commonly held misconceptions about climate change and global warming. Pass out Handout 4.5: Lines of Evidence and Handout 4.6: Explanations for Commonly Held Misconceptions About Climate Change and Global Warming.

2. Project or post these six commonly held misconceptions so the whole class can see them:

   - It's been warmer before.

   - Increases in solar activity are causing the Earth to warm.

   - The Arctic is gaining ice.

   - It hasn't warmed since 1998.

   - Increasing CO_2 has little or no effect.

   - Scientists don't agree that humans are causing the Earth to get warmer.

   Give students a few moments to review the misconceptions. Ask if anyone has heard any of these misconceptions and where they might have heard them. Don't spend too much time on the discussion—just enough to get their curiosity going.

3. Tell students you have provided them with some data sets that may help them understand why someone might hold onto these misconceptions. Tell them the data are from reliable sources such as NASA and NOAA and are found in Handout 4.5: Lines of Evidence. Encourage students to check these sources against Handout 4.1 just to be sure. They can work with classmates at their tables to understand the graphs. They may also use information in their notebooks from previous sessions to help them explain the misconceptions. Have students identify the data by graph number so you know which they are referring to.

4. Explain that students will work at their tables in groups but should create a way to record their information as individuals. Provide them with three categories to complete:

   - List the misconceptions.

   - Provide scientific reasoning revealing inaccuracies in each misconception.

- Explain why someone might accept the misconception.

   Students need to record the evidence that reveals the inaccuracies of the misconception and add their reasoning. Once they have explained why the misconception is inaccurate, they will use Handout 4.6 to decide why someone could agree with the misconception. They may choose more than one reason from their handouts.

5. Have students begin their work. Circulate around the classroom to clarify any questions about the graphs. In some cases, students may need to use the information from more than one line of evidence to dispel the misconception. Encourage students to use information from previous sessions in this unit.

Debrief

1. After about 10–15 minutes, bring the class together for discussion. Ask, "Is there a misconception you used to think might be accurate but now have evidence to show that it is not?" Have students explain their reasoning. Ask, "Which misconception was the easiest to invalidate? Why?" Ask, "Which misconception was the most difficult to invalidate? Why?"

2. Have students at each table share one misconception, their evidence and reasoning for why it is a misconception, and reasons they chose for why someone might agree with it. Some possible answers are listed in Table 4.1 (pp. 74–75).

3. Ask students if they noticed any patterns in the data. Give students a few minutes to look over the data and write their ideas in their notebooks. Have them talk with other students at their tables to compare ideas of what patterns they notice in any of the data sets. Once conversations wane, regain the attention of the class and ask volunteers to share any "ah-ha moments" they had during their discussions.

4. To sum up the session, ask, "Can you find evidence that global warming is *not* directly related to the consumption of fossil fuels by human beings?" Have students record their answers in their notebooks. Take as many comments from the class as you have time for.

5. It is recommended that you do not provide an opportunity for students to debate whether human-caused climate change is real. This can give students the inaccurate impression that there are two equal sides to the debate. Please see additional information in the Pedagogy section at the end of the session (p. 78).

Table 4.1: Understanding Misconceptions About Climate Change and Global Warming (Sample Student Work)		
Common Misconception	**Scientific Reasoning Revealing Inaccuracies in the Misconception**	**Explanation for the Misconception**
It's been warmer before.	Graph 4 in Handout 4.5 shows that there were four temperature peaks when it was warmer than our current temperatures. It looks like we were starting a cooling trend after the Holocene period. Looking at Graph 1 in Handout 4.5, it clearly shows that we were in a cooling trend until 1910 and now are in a significant warming trend. It's true that it's been warmer before, but there weren't people around at those times. It looks like people are causing it to get warmer this time when it should be getting colder.	1. A logical fallacy could lead you to think that the warming we are seeing now is normal, since it's been warmer before. 2. The data could be used to misrepresent a conclusion that may not be true.
Increases in solar activity are causing the Earth to warm.	Graph 5 in Handout 4.5 shows a variation in solar radiation, but it looks fairly even since 1979. There are no data to support an increase in solar radiation that could cause the Earth to warm like it is now.	1. I think this is a red herring because it is information that misleads you. 2. It could also be a logical fallacy because it makes sense when you first hear the misconception.
The Arctic is gaining ice.	Graph 2 in Handout 4.5 shows that Arctic ice has been declining steadily since about 1968.	1. Maybe there are fake experts who are saying the Arctic ice is increasing. 2. It could also be a red herring since it's not true at all.

Continued

Table 4.1 (continued)

Common Misconception	Scientific Reasoning Revealing Inaccuracies in the Misconception	Explanation for the Misconception
It hasn't warmed since 1998.	Graph 1 in Handout 4.5 shows that it's been warmer than average every year since 1977. However, it was extra warm in 1998 and took another seven years for the temperature to get that warm again. So, if you just look at the data during those seven years, you could come to the conclusion that it hasn't warmed since 1998, because it doesn't look like the temperature got warmer during this small chunk of time.	This is an example of cherry-picked data. If you consider all of the data until now rather than just select data points, you'll see that it has warmed a lot since 1998.
Increasing CO_2 has little or no effect.	When you look at the CO_2 levels from Graph 2 in Handout 4.2, you can see that they are currently much higher than at any other time in the last 400,000 years. This sharp increase began after the Industrial Revolution, when we started burning fossil fuels. When you compare that increase to the temperature increase in Graph 1 in Handout 4.5, it seems to reinforce that increased CO_2 makes the temperature increase.	I think a fake expert must have said this, since I can find no evidence to show it could be true.
Scientists don't agree that humans are causing the Earth to get warmer.	Graph 3 in Handout 4.5 shows seven studies that have been done on climate change and how many scientists agree with them. The lowest one had 91% agreement, and the highest one had 100% agreement. I think that most scientists agree that climate change is real.	1. Maybe this misconception is a conspiracy theory by oil companies. 2. This could be an example of jumping to conclusions if you didn't have all of the data.

Climate Change Agent Interview

1. Pass out Handout 4.7: Climate Change Agent Interview. The interview is with Dahr Jamail, a journalist promoting the truth about climate change. Give students a few minutes to read the interview, and then have them jot their reflections in their notebooks.

2. Ask students for any impressions they had of this interview. Ask if any part of the interview was difficult to read due to its honesty. Take a few responses. Ask, "If you experienced some of the things that Dahr Jamail has seen firsthand, like the Iraq war and the BP oil spill, how would you feel?" Ask if anyone disagrees with Jamail's opinions about our future world. Accept their disagreement. In this case, these are feelings, not facts. Therefore, they are not up for criticism or critique. Pose the question, "If you felt the same way as Dahr, what would you do?" This is meant to get students to reflect on the difficult nature of the truth about climate change and how difficult it might be to share this information with others.

3. Ask for any other comments before moving on to the statement review.

Reviewing Statements and Questions About Climate Change

1. Ask students to review the list they made in their notebooks about what they had heard about climate change. Ask if they now have enough evidence to determine the accuracy of any more of these ideas. If students have enough evidence to categorize a statement as accurate, they should record it on a sentence strip and post it in the Accepted as Accurate column. If they have enough evidence to categorize a statement as inaccurate, they should record it on a sentence strip and post it in the Accepted as Inaccurate column. If the evidence suggests that an idea on their list is accurate but not definitive, they may post it in the Needs More Information/Evidence/Research column. *In each case, students need to support their decision with evidence.*

2. Now revisit the statements posted during the previous sessions. Ask if anyone thinks a statement should be moved to a different column based on new information they learned during the current session. Students may decide that they need more evidence for a statement that they had previously placed in the Accepted as Accurate category, or they may now have enough evidence to move a statement that had been placed in the Needs More Information/Evidence/Research column to

either the Accepted as Accurate or Accepted as Inaccurate column. *In each case, they need to support their decision with evidence.*

3. Ask students if they can write new statements for any of the columns, based on new knowledge gained from this session. *In each case, they need to support their decision with evidence.*

4. Ask students to review the questions that are posted. Are there any questions that they can now answer with their new knowledge? If so, remove the question, write out the answer on a sentence strip, and post this statement in the Accepted as Accurate column. *In each case, they need to support their decision with evidence.*

5. Finally, ask students if they have any new questions to add to the Questions We Have About Climate Change column. Provide a few minutes for discussion and for posting questions.

Extending the Session

If you want students to meet the *NGSS* performance expectation HS-ESS2-4, include instruction on the natural variability of Earth's systems. One excellent resource for this is a tutorial at *www.sciencecourseware.org/eec/GlobalWarming/ Tutorials/Milankovitch*.

You could follow this with a simulation where students manipulate individual factors in the Milankovitch cycles to see how each affects global temperature. Access the simulation here: *https://cimss.ssec.wisc.edu/wxfest/Milankovitch/ earthorbit.html*.

Background for Teachers

Proxy Data

In Session 4, many of the data sets that students use are considered "proxy data." Proxy data do not come from direct measurement but rather from preserved physical characteristics of the environment that can stand in for direct measurements.

One example of proxy data is carbon dioxide bubbles trapped in ice cores over many thousands of years, which is studied during Session 4. Temperature values from thousands of years ago are determined by using proxy data. Paleoclimatologists extrapolate temperature from ratios of oxygen isotopes stored in sediments, tree rings, pollen, etc. The interview in Session 3 with Dr. Heidi Roop describes her work in this area.

A discussion of proxy data is appropriate if students ask how scientists know what the temperature was 300,000 years ago. You can get more information on proxy data here: *www.ncei.noaa.gov/news/what-are-proxy-data*.

Natural Cycles in Climate Change

Students may also raise questions about the role of natural cycles in climate change. The role of natural variation on climate variability was first described in 1920 by scientist Milutin Milankovitch. He is best known for developing one of the most significant theories about Earth's motion and long-term climate change.

He identified three factors and determined the relative influence of each:

1. **Eccentricity: The orbit of the Earth around the Sun varies from circular to slightly elliptical.** That means the Earth gets slightly closer to the Sun during peaks of eccentricity. These peaks occur every 95,000 years, but superimposed on those are larger peaks at 125,000 and 400,000 years.

2. **Obliquity: The tilt of the Earth's axis varies from 22.1° to 24.5°.** It is currently at 23.44°. The period of obliquity is 41,000 years.

3. **Precession: This is the Earth's slow wobble as it spins on its axis.** This toplike wobble, or precession, has a periodicity of 23,000 years.

Pedagogy

In Session 4, students will engage the science and engineering practice of argumentation. Argumentation is a key skill as it is the one practice that is consistent across all standards. This is a skill that overlaps with the *Common Core State Standards* and is developed heavily in both English language arts and mathematics.

For more depth in using scientific argumentation, these NSTA journal articles are excellent resources:

1. Blugren, J., and J. Ellis. 2015. Teacher's Toolkit: The argumentation and evaluation guide: Encouraging *NGSS*-based critical thinking. *Science Scope* 38 (7): 78–85.

2. Llewellyn, D. 2015. Teacher's Toolkit: Scaffolding students toward argumentation. *Science Scope* 39 (3): 76–81.

The "Debate" Over Climate Change

It is probable that discussion of the presumed controversy over human-caused climate change will arise in the classroom during this unit. It is important to remember that 97% of scientists worldwide agree that the causes of our rapidly warming planet are directly related to the burning of fossil fuels. To allow a

debate by "both sides of the argument" is akin to having a debate about geocentric versus heliocentric models of the solar system. Although a geocentric model was widely accepted in its day, overwhelming scientific evidence has shown that model to be inaccurate, and it has not been accepted for more than 200 years.

The same is true for climate change. As of this printing in 2019, the United States is the only country to withdraw from the Paris climate accord, which was signed by 195 countries. At the 21st Conference of the Parties (COP21) in 2015, every "party" (each country, plus the European Union) came together to set limits that would ensure a sustainable future for humans on Earth. Of all the parties at the conference, only two did not sign the agreement. One was Nicaragua, which claimed the limits did not go far enough. The other was Syria, which was steeped in a civil war. Since then, both Nicaragua and Syria have signed the agreement. In 2017, the United States committed to withdraw, taking effect in 2020. This should make any skeptic ask, "Why?" Every other country in the world agrees that climate change is real, is primarily caused by burning fossil fuels, and if unchecked could make the Earth uninhabitable to most humans. They also agree there are solutions that need to be put in place as soon as possible.

Assessment Opportunities

Student notebooks are designed to be used as assessment tools throughout the entire unit. As students now begin their research and presentations, it would be a good time to check the notebooks in order to assess misconceptions and knowledge gained at this juncture.

Resources

1. Students may enjoy playing a game that tests whether they can discern fake news from real news. They read a short article and decide whether it is real or fake. At the end of each round, they are scored based on how well they did. You can access the game here: *http://factitious.augamestudio.com.*

2. Alfonso Gonzalez, an exceptional science teacher at Chimacum Middle School in Chimacum, Washington, has put together a considerable number of resources on fake news, which you can find here: *www.diigo.com/profile/educatoral_?query=fake-news.*

3. Skeptical Science (*www.skepticalscience.com*) is an excellent resource with commonly accepted misconceptions about global warming and climate change along with the corresponding scientific explanations.

You can choose different levels of scientific explanations, from basic to intermediate to advanced.

4. A graphic is available from NASA (*http://climate.nasa.gov/evidence*) to help students see that the level of CO_2 rose markedly after the Industrial Revolution in the 1800s, when fossil fuels (coal, oil, natural gas, etc.) began to be burned at an increasing rate.

5. Students may be curious about the tilt and wobble of the Earth and its effect on climate. This interactive web page allows students to explore the concept of the Milankovitch cycles, the cause of the major ice ages of the past 500,000 years: *http://profhorn.meteor.wisc.edu/wxwise/climate/milankovich.html.*

References

North Carolina Climate Office. Milankovitch cycles. *https://climate.ncsu.edu/edu/Milankovitch* (accessed October 31, 2018).

Sustainable Innovation Forum. Find out more about COP21. *www.cop21paris.org* (accessed November 3, 2018).

SESSION 5

Conducting Research on Current Climate Change Topics

Introduction

This session leverages the power of student discovery, viewed through the lens of the *NGSS,* to support students' awareness of the complex nature of a warming planet.

Students are given the opportunity to research the many lines of evidence that support the overwhelming scientific conclusion that climate change is being fueled, literally, by human activity. Students form teams of experts, with each team focusing on one of the lines of evidence.

Working in groups of three, each student takes the lead on one element of the research project. This is designed to encourage individual student engagement and foster accountability among team members. Then students prepare a presentation that tells the story of their line of evidence, synthesizing information and evidence from their research.

The format is designed to teach students to behave like scientists, which means that their thinking is directed by the crosscutting concepts and their learning is directed by the science and engineering practices. As part of their analysis of information, students are tasked with developing computational models to predict the relationships between systems and their components in the natural and designed world. Students will organize their research into presentations that they will share at a "scientific conference" in Session 6.

Teachers also have the option of engaging students in independent investigations to model their chosen lines of evidence. These investigations give students a deeper understanding of their research topic and provide them with an authentic opportunity to use science and engineering practices to design experiments and interpret evidence. A list of possible investigations appears on pages 89–90 to help you guide students toward manageable projects. The results of their investigations can be incorporated into their presentations.

Objectives

1. To allow students to conduct their own information research on one line of evidence for climate change

2. To provide an opportunity for students to use discernment in conducting information research and choose sources that are accurate, respected, accountable, and verifiable

3. To challenge students to assimilate information from multiple sources and create a coherent and compelling presentation

4. To create structure for collaborative work in a group

5. To provide opportunities for students to conduct an investigation that supports the line of evidence they choose to research (optional)

What You Need

Gather the following materials.

For each group:

❑ Laptop or electronic device with internet access

❑ Copy of Handout 5.3: Scholarly Research Template

❑ Copy of Handout 5.4: Presentation Template

❑ Copy of Handout 5.5: Investigation Template (for optional independent investigations)

❑ Copy of Handout 5.6: Investigation Preplanning Template (for optional independent investigations)

Note: All handouts are found on the Extras page: www.nsta.org/climatechange.

For each student:

❑ Science notebook

❑ Copy of Handout 5.1: Climate Change Agent Interview

❑ Copy of Handout 5.2: Rubric for Climate Change Project

❏ Copy of Handout 5.7: Teamwork Evaluation for Students (optional group evaluation protocol)

Note: All handouts are found on the Extras page: www.nsta.org/climatechange.

Preparation

Before the Class

1. Adapt the templates in Handout 5.3 and Handout 5.4 to suit your students' abilities and your goals for this unit. You may want to add specific requirements for the students' research. The template in Handout 5.3 includes the term *positive and negative feedback,* which comes from the high school standards. Middle school teachers may want to replace this with the term *cause and effect.* The relevant *NGSS* performance expectations are listed below to assist you with choosing specific adaptations.

Suggestion for all students:

- Analyze data on natural hazards to forecast future catastrophic events and inform the development of technologies to mitigate their effects. (MS-ESS3-2)

For middle school students:

- Construct an argument supported by empirical evidence that changes to physical or biological components of an ecosystem affect populations. (MS-LS2-4)

For high school students:

- Feedback (negative or positive) can stabilize or destabilize a system. (Stability and Change)

- Use a model to describe how variations in the flow of energy into and out of Earth's systems result in changes in climate. (HS-ESS2-4)

2. Make one copy per student of the following:

   Handout 5.1: Climate Change Agent Interview

   Handout 5.2: Rubric for Climate Change Project

   Handout 5.7: Teamwork Evaluation for Students (optional group evaluation protocol)

3. For each group of three students, make one copy of the following:

   Handout 5.3: Scholarly Research Template

Handout 5.4: Presentation Template

Handout 5.5: Investigation Template (for optional independent investigations)

Handout 5.6: Investigation Preplanning Template (for optional independent investigations)

*(**Note:** Provide these as digital documents if document sharing is available for students.)*

4. Decide if you want students to read the Climate Change Agent interview in Handout 5.1 as homework or during class. If students are going to read it for homework, send copies of the interview home with them the day before you teach this session.

5. Decide how you would like to group your students—groups can be formed first and then each group can choose a topic to research, or students can individually pick a topic that interests them and form a group around that topic. Each group should have three members. They will be working closely together for the next few days. They may need to meet outside of school to complete their research and presentations.

6. A list of sample research topics is provided on pages 86–87. Students will be more engaged if they can choose a topic of high interest to them. If students choose topics not on the list, make sure that each one is sufficiently different from other topics being researched. Be prepared to coach students on their choices.

7. Each student will take on a specific role in their group:

   - **Project Coordinator:** This person will be responsible for assigning areas of research to each team member, ensuring validity of all sources, and overseeing the coordination of information between scholarly research and the presentation.

   - **Research Lead:** This person will be responsible for recording all information in Handout 5.3: Scholarly Research Template. They also need to ensure that all citations are recorded.

   - **Presentation Lead:** This person will be responsible for organizing the information from the research into an engaging and informative presentation for the class. They will use Handout 5.4: Presentation Template to guide their work.

8. Allow two regular class periods for research and one to two class periods for creating the presentation. If you need to shorten this time frame or if students need additional time, some tasks may be assigned as homework.

9. If you plan to assign the optional independent investigations, use the following division of responsibilities and timeline rather than those listed above in Steps 7 and 8:

- **Scholarly Research Lead:** This person will take the lead on information research. They will be responsible for assigning areas of research to other team members, recording information in Handout 5.3: Scholarly Research Template, and ensuring that all sources are valid.

- **Investigation Lead:** This person will be responsible for directing the work on the investigation, recording all investigation design information and data, and completing data analysis using Handout 5.5: Investigation Template.

- **Presentation Lead:** The third person will be responsible for coordinating the information from both the research and the investigation to create the class presentation, using Handout 5.4: Presentation Template to guide their work.

  This adjusted schedule requires additional time for the investigations: Allow two regular class periods for research, one class period for selecting and designing the investigations, one to two days for carrying out the investigations, one day for analyzing the investigations, and one day for preparing the presentations.

Begin!

Preparing for Research

1. If you did not assign the Climate Change Agent interview as homework, distribute Handout 5.1 at this time. Give students five minutes to read it and make comments in their notebooks. The interview is with Dr. Robert Bindschadler, a scientist who studied glaciers and ice sheets for 30 years in Antarctica. He has a glacier and an ice stream named after him!

2. Ask students to think about the interview with Dr. Bindschadler and give their impressions. In addition to describing the work he did for 30 years with NASA in Antarctica, Dr. Bindschadler makes a number of points in his interview about the nature of science and what scientists do. Explain to students that they will be conducting research about topics that contain information from scientists around the world.

3. Draw students' attention to the current Questions We Have About Climate Change list. Ask them to review these questions and decide

which ones interest them the most. Have them list or highlight those questions in their notebooks. After a few minutes, have them look over the accurate and inaccurate statements that have piqued their interest and highlight those in their notebooks, as well.

4. Explain that students will be working in research teams of three to study one aspect or line of evidence for climate change. They will be acting like scientists in conducting information research to find out more about climate change. In Session 6, students will present their information at a mock scientific conference on climate change, just as scientists do.

5. Give students 10–15 minutes to think about the topics that interest them and review the information in their notebooks. Ask each student to identify one piece of evidence or a concept that seems significant to them.

6. After they have had time to review this information, have students share their thinking with others at their table. They should record the significant concepts or evidence from each group member in their notebooks. This may help guide their topic selection.

7. Have them talk at their tables or in pairs to decide what "big ideas" they should keep in mind when conducting their research. (For instance, students should think about CO_2 emissions in the United States and worldwide, sources of greenhouse gases, population projections, etc.)

Choosing a Research Topic

1. Distribute Handout 5.2: Rubric for Climate Change Project, and allow time for clarifying questions. This rubric is for assessment of knowledge and overall quality of work. Explain that students will be evaluated based on how well they communicate what they learned from their research. Students should use the rubric as they work together on their projects so they know what is expected. Students will also be asked to reflect on their individual participation and the effectiveness of their group.

   If you plan to use the optional Handout 5.7: Teamwork Evaluation for Students, distribute it at this time, as well.

2. Provide students with a list of potential research topics, or leave it open for them to choose what interests them most. A list of sample topics is included below. Make sure each group has a different topic to research.

   Sample Research Topics:

   • Ocean acidification

- Sea level rise
- Declining Arctic sea ice/albedo effect
- Extreme weather events
- Global temperature rise
- Warming oceans
- Shrinking ice sheets in Greenland and Antarctica
- Glacial retreat
- Increase in diseases; changes in distribution of disease
- Deforestation through logging and/or burning
- Disruption of migration patterns
- Thawing of permafrost and methane release

3. Complete project requirements are provided in Handout 5.3: Scholarly Research Template and Handout 5.4: Presentation Template. Allow students to review the templates and select the element of the project that they will lead. Be clear that although they are the lead on one element, all team members must collaborate and share information as they work. Remind students that they will be required to cite their research from reputable and respected sources.

Scholarly Research

1. Students will need access to internet sources. Provide two days for students to research as a team, or have them conduct research one day in class and one night for homework. Allow students to engage deeply in their research to gain an understanding of their topic before creating a presentation.

2. The Project Coordinator will direct the work of the group. This person assigns areas of research to each person and ensures that sources are valid. All students participate in the research, but the Research Lead is responsible for completion of the template. Using a digitally shared document makes it easy for students to work synchronously.

3. As students identify sources, they need to use the guidelines for valid research established in Session 4. All sources need to be recorded in Handout 5.3: Scholarly Research Template. Use the style sheet established by your school for recording citations.

4. Research not completed during the allotted class time should be completed as homework.

Preparing a Presentation

1. Remind students that each group has been researching a different topic or line of evidence that relates to climate change. It is important that they share their information in a way that is accurate, interesting, and compelling. Each group will be sharing their topic in a five- to seven-minute presentation during a mock climate change conference in Session 6. This is their opportunity to prepare.

2. Students may use PowerPoint, Prezi, or another presentation format that is supported by technology at your school. Handout 5.4: Presentation Template serves as a guide to ensure that all necessary information is included.

3. The Presentation Lead is in charge of transferring appropriate research on their topic into the Presentation Template. They will need to work collaboratively with the other team members to complete the task.

4. Explain that students will need to answer questions about the validity of their information and provide reliable sources for all evidence used as they share their presentations in Session 6.

Reflection

As students finish this session, provide an opportunity for reflection. Ask them to record answers to the following questions in their notebooks. (After students answer the questions, you can have them complete the optional assessment from Handout 5.7: Teamwork Evaluation for Students, if you choose.)

1. Describe the work that you contributed to the group process. Why was this work important to developing group understanding of your topic?

2. What problems did you encounter in your personal work and in your group work? How did you resolve those problems?

3. In what ways was this division of tasks an effective strategy to ensure that all group members participated and communicated in a meaningful way? How could you improve this process?

4. What was the most interesting or important concept that you learned through the research of your topic?

5. What proportion of the total work done by your group did you complete? (For members in a group of three to complete an equal amount of work, each person must do one-third of the total work.) Support your claim with evidence.

Optional Independent Investigations to Support Research

This option prompts students to use *NGSS* skills to design and carry out an independent investigation that models the line of evidence they researched. It is challenging to gather the diverse materials and equipment needed for 8 to 10 different investigations occurring at once. However, the investigations provide a very rewarding experience for both the teacher and students. Should you decide to include the investigations in the session, students can provide many of the required materials.

1. As students finish the research element of the project, challenge them to design their own investigation to model the system they studied.

2. As you circulate around the classroom, ask leading questions to help direct students to reasonable investigations based on available equipment and materials. This challenges students to apply their knowledge of scientific design to create valid tests, using controls and repetitions. If students are having difficulty finding or creating investigations, a list of suggestions is provided here.

Sample Investigations to Support Research Topics:

*   **Ocean acidification:** Determine the percent change in mass of oysters, clamshells, or chalk in varying acid solutions. Vinegar works well for this. For a complete description of this experiment, refer to the article "Ocean Acidification," published in the September 2015 issue of *The Science Teacher*. (See the Resources for Student Research section, p. 95, for more information.)

*   **Sea level rise:** Use a slanted paint tray to measure the increase in land covered by water compared to the rise in the depth of the water.

*   **Declining Arctic sea ice/albedo effect:** Measure changes in the melt rate of regular ice versus ice covered with a light coating of pencil-sharpener shavings when exposed to a heat source.

*   **Extreme weather events:** Create a cloud in a 1-liter plastic bottle. Assess change in the cloud formation with different water temperatures. Students can find directions for this online.

*   **Global temperature rise:** Create a miniature greenhouse using plastic wrap–covered boxes and a light source to model the impact of the greenhouse effect on temperature. Use different thicknesses of plastic to cover the boxes, or line the bottom of the boxes with different types of substrate.

*   **Warming ocean:** Measure the increase in volume of water as the temperature rises. Narrow-neck volumetric flasks work best for

this. For a complete description of this experiment, refer to the article "A Rising Tide," published in the September 2015 issue of *The Science Teacher*. (See the Resources for Student Research section, p. 95, for more information.) Another option would be to monitor the heart rate of Daphnia in different water temperatures.

- **Shrinking ice sheets in Greenland and Antarctica:** Design an investigation to answer one of the following questions: What causes ice to melt the fastest—heat applied to the top surface of the ice or warmed water under the ice? How does surface area affect the melt rate of ice?

- **Glacial retreat:** Fill one beaker with water and ice cubes to a set volume. Add the same volume of water to a second beaker, but put the ice on a screen placed over the beaker so the ice drips into the beaker. Measure changes in volume as the ice melts.

- **Increase in diseases; changes in distribution of disease:** Grow bread yeast at different temperatures for 24 hours, and then use a microscope or probeware to determine the concentration of yeast. Alternatively, raise mealworms at different temperatures to model the change in the pupation rate of insects (potential vectors of disease) as climate warms. This will take five to seven days to complete.

- **Deforestation through logging and/or burning:** Use paint trays as stream tables. Fill two trays with sand. In one stream table, add small twigs, bits of grass, and other objects stuck into the sand to represent vegetation. The other stream table will only have sand. Pour a controlled amount of water in each and record differences in patterns of erosion. Another option is to use probeware to measure the carbon dioxide and oxygen levels in biochambers with living plants versus bare soil.

- **Disruption of migration patterns:** Germinate either Wisconsin fast plant seeds or radish seeds at different temperatures to model changes in plant germination rates, which may affect food availability for migrating animals. Wisconsin fast plant seeds germinate in one day at room temperature. Premoistened radish seeds take two days to germinate.

- **Thawing of permafrost and methane release:** This one is tricky. Permafrost stores thousands of years' worth of organic matter. As it thaws, decomposition speeds up and the gases of decomposition are released. Since the conditions of permafrost are largely anaerobic, methane is released as it decomposes. One way to model this is to peel a banana and divide it into three equal pieces.

Roll all of the banana pieces in yeast, and then place each one in a separate resealable bag. Smash each banana, and then squeeze all of the air out of its bag before sealing. Store each bag at a different temperature (refrigerator, room temperature, incubator). Students should notice some difference during the class period, with more dramatic results if left overnight.

Note: The following topics have the potential for complex interactions that require a more nuanced approach and may not easily lead to student-generated investigations. Consider the ability levels of your students as you guide them toward topic selection.

- Increase in diseases; changes in distribution of disease

- Disruption of migration patterns

- Thawing of permafrost and methane release

3. Students will use Handout 5.5: Investigation Template to complete all aspects of the investigation. The Investigation Lead will transfer pertinent information from Handout 5.3: Scholarly Research Template to the Background Information portion of the Investigation Template.

4. Once students have decided on their investigation, have them complete Handout 5.6: Investigation Preplanning Template. You will use the information on this to ensure that investigations are reasonable and that necessary materials and equipment are gathered for the appointed investigation day(s).

Safety Note: Ensure that students have proper safety equipment needed to carry out their investigations. Students should wear safety goggles, aprons, and nitrile gloves as appropriate for their investigation. Remind them to use caution when working with glassware. Wear heat-resistant gloves when heating glassware. Never heat a closed system. Any breakage should be cleared by the teacher. All spills should immediately be wiped up as this is a slip/fall hazard.

5. Student groups will be working independently on the day of the investigation. The members of each group all participate in setting up and carrying out their investigation, recording data, and cleaning up once they've finished.

6. If students realize that their controls are not adequate, allow them to modify and improve their procedures.

7. Some investigations require more than one day. The Investigation Lead can finish data collection while others in the group are working on the presentation. Alternatively, students can come in before or after school to gather final data.

8. Provide a class period for data analysis and completing the Investigation Template. As individuals and groups finish this work, they may move into preparing for their presentation.

9. A video of students describing their integrated research and investigations is provided for you here: *www.youtube.com/watch?v=TaVyv7XUjOA&fetaure=youtu.be.*

Background for Teachers

Pedagogy

Session 5 is made up of identified tasks. This division requires team members to work collaboratively and allows each student to bring different skills and abilities to the team. Rather than dividing each task equally (e.g., having each person research three sources and create four slides for the PowerPoint), you will have each student focus on one element of the project. This provides opportunities to differentiate instruction for students who excel in graphic design, mathematical reasoning, organization, and technical problem solving. It also encourages students with different skill sets to work together for maximum benefit.

Students may be overwhelmed with the dire warnings about the consequences of a warming Earth and the changes that are predicted. It is important to balance their concerns with opportunities to research viable solutions and evaluate their potential effect. (Students will investigate solutions more deeply in Session 8 and Session 9.) Investigating solutions also provides another opportunity for using models to predict positive future change.

The optional independent investigations of Session 5 provide an opportunity for students to design and carry out an investigation that models the line of evidence they are researching. Students will apply their ability to use models to make predictions based on original data. This builds deep understanding of their topic. Some topics lend themselves to sophisticated models, whereas others can be quite simple. Consider this as you guide students toward their topics if you choose to incorporate the optional investigations.

If your students do not have experience designing their own investigations, you may need to scaffold this with some whole-class instruction. Make sure you focus on safe lab procedures and proper disposal of materials. The designated day of investigation is a challenging but exciting day for the teacher and students. Each group will be setting up a different lab. They may encounter unexpected needs as they work. Allowing for a bit of trial and error is challenging for the teacher, but it is where students really grow to understand the relationship between independent and dependent variables, controlling variables to ensure validity, and planning repetition to ensure reliability. You will find that while

your lab is busy and noisy, this is a time of maximum engagement and learning. Sometimes investigations yield unexpected results. Use this as a learning opportunity. If time is available, allow students to redesign their investigations and start over. This provides an excellent example of the struggles that scientists have when designing investigations to test new ideas.

Assessment Opportunities

There are three opportunities for assessment of student thinking and participation in this session.

- Handout 5.2 features a comprehensive rubric that students use to assess their presentations on the basis of content, communication of information, completeness, organization/clarity, documentation, and grammar/mechanics. In Session 6, each student will complete this rubric as a final assessment of their presentation.

- Handout 5.7: Teamwork Evaluation for Students is an optional assessment that evaluates individual participation in the group presentations. Students use the handout to assess their teammates' contributions, assigning a fictional monetary payment to each team member based on the amount of work they did.

- The Reflection section on page 88 includes five questions that provide an opportunity for students to reflect on their role in the group and on new understandings they gained through their project. As students record their thinking in their notebooks, they provide insight into their growing understanding of the complex topic of climate change. The student notebooks are designed as an ongoing assessment of student thinking to be used throughout the entire unit.

Resources for Student Research

1. NOAA Climate.gov (*www.climate.gov/#climateWatch*) provides *lots* of information, interactives, visual data, and more.

2. The NASA Global Climate Change website includes a Facts page (*http://climate.nasa.gov/evidence*), which features informative content. The Evidence section of the web page begins with the critical "hockey stick" graph of CO_2 levels from the past 400,000 years. The web page also includes sections on the causes of climate change, the effects of climate change, and more. Be sure to check out the Interactives section on the site's Explore web page, as well.

3. The Environmental Protection Agency (EPA) partners with more than 40 data contributors from various government agencies, academic institutions, and other organizations to compile a key set of indicators related to the causes and effects of climate change. The indicators are published in this report, available on the EPA's website and in print: *www.epa.gov/climate-indicators.*

4. Earth Vision Institute has created Getting the Picture: Our Changing Climate, an innovative online multimedia tool for climate education (*http://gettingthepicture.info*). Here, acclaimed photographer James Balog and his team provide the latest in climate science education, featuring unique archives of media, film, photography, and firsthand accounts of our changing climate. (Balog is the featured Climate Change Agent in Session 7.)

5. Check out research from the National Audubon Society on the changes in migratory patterns of birds and their population fluctuations: *http://web4.audubon.org/bird/bacc/Species.html.*

6. With NOAA's Data in the Classroom, students use real-time ocean data to explore today's most pressing environmental issues and develop problem-solving skills employed by scientists. Go to this site to access online and classroom-ready curriculum activities with a scaled approach to learning and easy-to-use data exploration tools: *http://dataintheclassroom.org.*

7. Budburst is a citizen science program focused on understanding plant phenophase timing and its response to environmental change. You and your students can join the national Budburst network of scientists, students, teachers, and volunteers and monitor plants as the seasons change to help the organization understand how plants respond to climate change. Visit this site for more information: *https://budburst.org/educators.*

8. The Fourth National Climate Assessment is a report that summarizes the current and future impact of climate change on the United States. A team of more than 300 experts guided by a 60-member federal advisory committee produced the report, which was extensively reviewed by both the public and other experts, including federal agencies and a panel from the National Academy of Sciences. Get the report here: *https://nca2018.globalchange.gov/chapter/front-matter-guide.*

9. The Climate Literacy and Energy Awareness Network is a collection of more than 700 free, ready-to-use resources rigorously reviewed by educators and scientists. The resources can be selected by grade level, by resource type (activity, short demonstration/experiment, teaching

guidance, video, or visualization), and by *NGSS* topics. You can access the collection here: *https://cleanet.org/index.html*.

10. You may find these two resources from the National Academies of Sciences, Engineering, and Medicine helpful in this session:

   • Download a free booklet from the organization on climate change evidence and causes: *http://nas-sites.org/americasclimatechoices*.

   • The following web page is dedicated to creating, evaluating, and understanding models of climate change: *http://nas-sites.org/climate-change/climatemodeling*.

11. The Global Learning and Observations to Benefit the Environment (GLOBE) Program is an international science and education program that provides students and the public with the opportunity to participate in data collection and the scientific process and to contribute meaningfully to our understanding of the Earth's system and global environment. This student-centered page includes links to student research reports from around the world. Students can also learn how to be part of the GLOBE research, play games to learn about the Earth as a system, and more. Visit this site for more information: *www.globe.gov/do-globe/for-students*.

12. For a complete description of the experiment to demonstrate the effect of ocean acidification, refer to the following article: Ludwig, C. et al. 2015. Ocean acidification. *The Science Teacher* 82 (6): 41–45.

13. For a complete description of the experiment to demonstrate the effect of a warming ocean on water volume, refer to the following article: Trendall M., A. Feldman, and P. Wang. 2015. A rising tide. *The Science Teacher* 82 (6): 34–40.

14. Students studying the thawing of permafrost may be interested in the video "Exploding Methane Gas Bubbles" from the BBC series *Earth: The Power of the Planet*. The video features scientists drilling into a frozen lake to ignite methane gas trapped in bubbles beneath the surface. To watch the video, visit this site: *www.science.org.au/curious/video/exploding-methane-gas-bubbles*.

SESSION 6

Climate Change Conference

Introduction

After conducting their research and creating their presentations, students engage in a mock scientific conference to demonstrate another aspect of the work of scientists—sharing work with peers and seeking critical review. To set the stage for their conference, students learn about an international event called Conferences of the Parties (COP), which is held annually to address climate change on the world stage.

The class will simulate a COP where elite scientists from around the world present their best research on topics involving climate change. You may want to invite community partners or administrators to be part of the audience to increase the sense of importance for high-quality work.

Peer review is an essential component of the scientific process. It highlights the importance of critical listening. As one group presents, others listen carefully so they can ask questions that challenge and encourage student thinking.

Finally, students self-evaluate using the rubric they were given in Session 5 to determine if they have reached the learning targets. The teacher will be using the same rubric to assess student learning.

After the conference, students return to the list of questions generated in Session 1. They may now be able to answer a large number of them with all of the new information they have gained during the conference.

 **Session 6**

Objectives

1. To provide an opportunity for students to present their research in a format that conveys the information effectively to their fellow classmates

2. To simulate a science conference similar to the annual international Conference of the Parties

3. To provide peer-reviewed feedback to each team of students on their presentations

4. To provide a process for engaging all students in understanding a range of topics on the complex nature of climate change

What You Need

Gather the following materials.

For the class:
- ❑ Display area for posters
- ❑ Laptop or electronic device for projecting student presentations
- ❑ Digital projector
- ❑ Other multimedia support as required for presentations
- ❑ The current list of questions and statements on the wall OR access to the electronic documents with questions and statements
- ❑ Sentence strips for additional statements (if you're using the wall columns)
- ❑ Marking pens (if you're using the wall columns)

For each group:
- ❑ Peer review form from Handout 6.1: Climate Change Presentation Peer Review (1 for each presentation observed)
- ❑ Copy of Handout 6.2: Sentence Stems (1 half-sheet per group)

For the teacher to complete for each presenting team:
- ❑ Copy of Handout 5.2: Rubric for Climate Change Project (from Session 5)

For each student:

❑ Science notebook

❑ Copy of Handout 6.3: Climate Change Agent Interview

Note: All handouts are found on the Extras page: www.nsta.org/climatechange.

Preparation

Before the Class

1. Be sure to let students know the time frame for presentations as early as possible in their research. It is recommended that the presentations last between five and seven minutes.

2. Check in with each team to see what they will need in terms of materials or multimedia equipment and support.

3. Decide what order the teams will take for their presentations, and make each aware of the assigned time slot.

4. When scheduling the presentations, allow for ample time at the end of each presentation (around 10–15 minutes) for

   - questions from the class;

   - students to review their notebooks to revise any misconceptions and add new knowledge from the presentation;

   - presenting teams to complete an evaluation of their presentation using Handout 5.2: Rubric for Climate Change Project (from Session 5);

   - students to complete the team presentation evaluations from Handout 6.1;

   - you to use the rubric to evaluate each team's presentation; and

   - a discussion about any statements and questions that need to be moved to a different column on the wall.

5. Make copies of Handout 6.1: Climate Change Presentation Peer Review. These must be double-sided copies since both sides are needed for the review. Cut these sheets into quarters along the lines provided. Each group will need one quarter-sheet for every other group's presentation. For example, if there are eight presentations in all, each group will need to evaluate seven other groups and therefore need seven double-sided copies of the peer review form.

6. Make copies of Handout 6.2: Sentence Stems, with each sheet cut in half. One half-sheet is used per table. You may want to laminate these so they can be reused.

Begin!

Setting the Stage for a Conference on Climate Change

1. Explain to students that when scientists conduct research, they often present their findings at conferences. Some of their presentations are peer-reviewed by fellow scientists in the audience. They will be participating in a similar format at this conference.

2. There have been several notable conferences on climate change in recent years. Mention these to your students, and have them add any information they may have gathered during their research.

Kyoto, Japan, 1997: The Kyoto conference on climate change took place at the Conference of the Parties III (COP3) to the United Nations Framework Convention on Climate Change. There, developed countries agreed to specific targets for cutting their emissions of greenhouse gases. A general framework was defined for this, with specifics to be detailed over the next few years. This became known as the Kyoto Protocol. Thirty-six countries, including all members of the European Union, have already signed the Kyoto Protocol. The United States has not.

Bali, Indonesia, 2007: Touted as the largest conference on global climate change in history, 190 countries were represented. The United States sent more than 100 delegates. All 27 countries that made up the European Union at the time sent national teams.

Copenhagen, Denmark, 2009: The Copenhagen climate change conference (called COP15) raised climate change policy to the highest political level. Close to 115 world leaders attended the high-level segment, making it one of the largest gatherings of world leaders ever outside UN headquarters in New York City. More than 40,000 people, representing governments, nongovernmental organizations, intergovernmental organizations, faith-based organizations, media, and UN agencies, applied for accreditation.

COP15 was a crucial event in the negotiating process:

– It significantly advanced the negotiations on the infrastructure needed for effective global climate change cooperation, including improvements to the Clean Development Mechanism of the Kyoto Protocol.

- It produced the Copenhagen Accord, which expressed a clear political intent to constrain carbon and respond to climate change in both the short and long term.

The Copenhagen Accord contained several key elements on which there was a strong convergence of the views of governments. This included the long-term goal of limiting the maximum global average temperature increase to no more than 2°C above preindustrial levels, subject to a review in 2015. There was, however, no agreement on how to do this in practical terms.

Paris, France, 2015: The 21st Conference of the Parties (COP21) made global history! As a result of COP21, 196 countries adopted the Paris Agreement, a new legally binding framework for an internationally coordinated effort to tackle climate change. The agreement represents the culmination of six years of international climate change negotiations and was reached under intense international pressure to avoid a repeat failure of the Copenhagen conference in 2009. Every "party" (each country, plus the European Union) came together to set limits that would ensure a sustainable future for humans on Earth. Of all the parties at the conference, only two did not sign the agreement. One was Nicaragua, which claimed the limits did not go far enough. The other was Syria, which was steeped in a civil war. Since then, both Nicaragua and Syria have signed the agreement.

The agreement establishes a global warming goal of well below 2°C above preindustrial averages. It requires countries to formulate more ambitious climate targets that are consistent with this goal. For the first time, all countries will develop plans on how to contribute to climate change mitigation and communicate their "nationally determined contributions" to the Secretariat of the Convention every five years. A coalition of more than 100 countries successfully lobbied not only for a lower temperature goal but also for a legally binding agreement, a mechanism for reviewing countries' emissions commitments every five years, and a system for tracking countries' progress toward meeting their mitigation goals.

In 2017, the United States was the only country to announce its withdrawal from the Paris climate accord, taking effect in 2020.

Much more is needed. The Paris Agreement is an important first step toward an effective policy where ambitious measures are taken to keep our emissions low enough to maintain no more than 2°C of warming, to ensure transparency of measures taken by each country, and to review our progress toward that goal every five years.

For more information on these conferences, students may use the following resources.

- Watch a short video to summarize COP21: *www.youtube.com/watch?v=I-4F5MJEeqs*.

- A list of all COP meetings from 1995 to present can be found here: *https://en.wikipedia.org/wiki/United_Nations_Climate_Change_conference*.

Attending the Conference on Climate Change

Setting Up the Conference

1. Describe the format you have chosen for the presentations, including the order in which teams will present.

2. Distribute copies of Handout 6.1: Climate Change Presentation Peer Review so that each group has a review form for every team that is presenting. Also give a copy of Handout 6.2: Sentence Stems to each group. Emphasize to students that they are to focus on the content of each presentation when filling out their reviews. These evaluations will be given to the teams to read.

3. Remind students that they are participating in a conference, just as scientists do. As such, they will behave as scientists and show respect to their fellow presenters.

Begin the Presentations

1. After each group has given its presentation, allow time for the students who were listening to form questions. They may use the sentence stems in Handout 6.2 to guide their questions or come up with their own. Students should record their questions in their notebooks. Encourage each group to propose at least one question to the presenters.

2. When you feel that there has been adequate time for questions and answers, students in the presenting group may return to their seats. All other groups will collaboratively complete Handout 6.1: Climate Change Presentation Peer Review. Ask them to focus on the quality of the content, science practices, and reasoning as they provide feedback.

3. While the groups are discussing and completing their peer review, the students who presented should complete their personal assessment using Handout 5.2: Rubric for Climate Change Project. In the space provided, they should support the scores they selected in each category with evidence from their work.

4. Finally, ask all students to record the most important concepts that they learned from the presentation and how they relate to their own topic.

While students are writing in their notebooks, queue the presentation for the next group.

5. Collect the peer-review forms from each group. Quickly scan them to ensure they are appropriate to share. When all presentations are finished, hand the feedback forms to the presenting groups. Re-collect them after students have finished reading them, and use them as part of your evaluation of their work.

6. Collect the following from each presenting group:

   • Handout 5.2: Rubric for Climate Change Project (one from each student)

   • Handout 5.3: Scholarly Research Template, Handout 5.4: Presentation Template, and if used, Handout 5.5: Investigation Template; these may be shared digitally.

7. At the end of each COP, the nations involved sign agreements to address the consequences of climate change based on the evidence shared in all of the presentations. Students will make pledges of changes they perosnally want to make at the end of Session 9.

Climate Change Agent Interview

1. Pass out Handout 6.3: Climate Change Agent Interview. The interview profiles Kate Chadwick, who works as an analyst with the World Bank Group. Give students a few minutes to read the interview, and then have them jot down their reflections in their notebooks.

2. Ask students for any impressions they had from this interview. Kate Chadwick has played a significant role in international climate change policy. Have students imagine what it must be like to work with her. If students could ask her a question, what might it be? They can jot down these questions in their notebooks.

3. Ask for any other comments before moving on to the statement review.

Reviewing Statements and Questions About Climate Change

1. Ask students to review the list they made in their notebooks about what they had heard about climate change. Ask if they now have enough evidence to determine the accuracy of any more of these ideas. If students have enough evidence to categorize a statement as accurate,

they should record it on a sentence strip and post it in the Accepted as Accurate column. If they have enough evidence to categorize a statement as inaccurate, they should record it on a sentence strip and post it in the Accepted as Inaccurate column. If the evidence suggests that an idea on their list is accurate but not definitive, they may post it in the Needs More Information/Evidence/Research column. *In each case, students need to support their decision with evidence.*

2. Revisit the statements posted during the previous sessions. Ask if anyone thinks a statement should be moved to a different column based on new information they learned during the current session. Students may decide that they need more evidence for a statement that they had previously placed in the Accepted as Accurate category, or they may now have enough evidence to move a statement that had been placed in the Needs More Information/Evidence/Research column to either the Accepted as Accurate or Accepted as Inaccurate column. *In each case, they need to support their decision with evidence.*

3. Ask students if they can write new statements for any of the columns, based on new knowledge gained from this session. *In each case, they need to support their decision with evidence.*

4. Ask students to review the questions that are posted. Are there any questions that they can now answer with their new knowledge? If so, remove the question, write out the answer on a sentence strip, and post this statement in the Accepted as Accurate column. *In each case, they need to support their decision with evidence.*

5. Finally, ask students if they have any new questions to add to the Questions We Have About Climate Change column. Provide a few minutes for discussion and for posting questions.

Extending the Session

C-ROADS (*www.climateinteractive.org/tools/c-roads*) is an excellent simulation of the type of negotiations that occurred at COP21 in Paris. It is a free, award-winning computer simulator that helps students and adults understand the long-term climate impacts of actions that reduce greenhouse gas emissions. Students take on the role of different countries and lobby one another to limit and draw down their CO_2 output, encourage reforestation, and discourage deforestation. You can use C-ROADS to rapidly test strategies for tackling climate change and watch the graph of temperature and CO_2 change as you input data. C-ROADS World Climate is available for download to Windows and Mac and runs online. This activity can be run in one class period if you do just one round of negotiations or in two periods if you allow the second round of negotiations.

Background for Teachers

Pedagogy

Encourage students to think about their audience when developing visuals for the presentation and figuring out how to tell the story of their topic. They can reflect back to the interview with Dr. Robert Bindschadler in Session 5, where he states that scientists have to be good at telling their story.

In this session, active listening is as important as presenting. There are many lines of evidence that support the theory of human-caused climate change. Each group is only researching one line of evidence. To get the whole story, students need to be active learners during each presentation. To encourage this, engage students in actively forming questions while they listen to each presentation. Sometimes it is difficult for students to know how to begin forming questions. Several sentence stems are provided to guide students in forming their questions. You may want to select three to five sentence stems that you think will be most effective with your students and their experience. The goal is for students to respectfully receive critiques and challenge others' ideas and conclusions of others using scientific reasoning.

A powerful way for students to share their research is to put on a presentation for the public. Plan for an evening event where parents and community members are invited, with a scheduled keynote speaker followed by students presenting in concurrent sessions. A sample program from Earth Summit, a public climate change presentation held at Port Townsend High School, is provided for you here: *https://docs.google.com/document/d/1h1twCYz55ZkXzKmd2og7mgd 97cxuMcgmXJbTudcDX2k/edit?usp=sharing.* (Note that student names have been removed from this sample publication, but these names are usually included in the program.)

Be sure that you provide an opportunity for students to reflect on their personal growth in understanding as they consider the connections between the various lines of evidence presented by their peers. "Reflection is essential if students are to become aware of themselves as competent and confident learners and doers in the realms of science and engineering" (NGSS Lead States 2013, p. 66).

Assessment Opportunities

Following each presentation, members of the presenting group will complete their self-evaluations in Handout 5.2: Rubric for Climate Change Project, including evidence from their work to support their scores. The teacher will also complete a rubric for each team, and then compare responses. If you notice

significant differences between your assessment and theirs, use this as an opportunity for discussion. If a student score is higher than yours, allow them to make their case for the higher score, using evidence from their work. Students are usually honest but can be quite critical of their own work. Find opportunities to recognize quality and promote student growth where students undervalue their contributions to the project.

You will also want to review students' notebook entries for this session, looking for their ability to make connections between complex ideas and ask meaningful questions.

References

Van der Gaast, W., and M. Alessi. 2015. History of the UN climate negotiations, Part 1: From the 1980s to 2010. Climate Policy Info Hub. *http://climatepolicyinfohub.eu/history-un-climate-negotiations-part-1-1980s-2010.*

NGSS Lead States. 2013. *Next Generation Science Standards: For states, by states.* Washington, DC: National Academies Press. *www.nextgenscience.org/next-generation-science-standards.*

UN Climate Change. Process and meetings: Conference of the Parties (COP). United Nations. *https://unfccc.int/process/bodies/supreme-bodies/conference-of-the-parties-cop* (accessed February 4, 2019).

SESSION 7

Climate Change Challenges

Introduction

This session provides an opportunity for students to integrate the information they have learned from their research. Students create charts or mind maps to anticipate the ripple effects from an aspect they have studied. This encourages students to think deeply about the consequential effects of a warming climate. For example, if students studied sea level rise, they will not only consider a predicted rise of a foot or more of sea level but also consider the secondary effect of that water flooding the downtown area of a low-lying coastal city or town. They will then consider the tertiary effects of flooding in the downtown area in terms of the environmental, human, and economic impacts. Through this process, students will have a clearer understanding of the wide-ranging effects these predicted changes will have on our communities. It promotes a type of systems thinking that goes beyond facts and concepts to a broader understanding of the issues.

The session culminates with students choosing a prominent "effect pathway" to share with their classmates, providing a point of emphasis from their research. If these ripple effects are not drawn out in the students' research presentations in Session 6, they are demonstrated to the class in this session.

Objectives

1. To introduce the concept of a chain of effects that can result when a seemingly small change is made to a complex system

2. To provide students with an opportunity to explore a range of primary, secondary, and tertiary effects that may result from one of the climate change topics presented at their conference

3. To integrate the knowledge gained in Session 6 and apply it to consequential thinking of climate change effects

What You Need

Gather the following materials.

For the class:

❑ The current list of questions and statements on the wall OR access to the electronic documents with questions and statements

❑ Sentence strips for additional statements (if you're using the wall columns)

❑ Marking pens (if you're using the wall columns)

For each group of four students:

❑ Laptop or electronic device with internet access

For each student:

❑ Science notebook

❑ Copy of Handout 7.1: Climate Change Effects Wheel or other chart design you provide for them (optional)

❑ Copy of Handout 7.2: Climate Change Agent Interview

Note: All handouts are located on the Extras page: www.nsta.org/climatechange.

Preparation

Before the Class

1. Look over Figure 7.1 and Figure 7.2. They show ways in which students can organize the information they collect on the effects of climate change. If you would like to provide options for differentiated

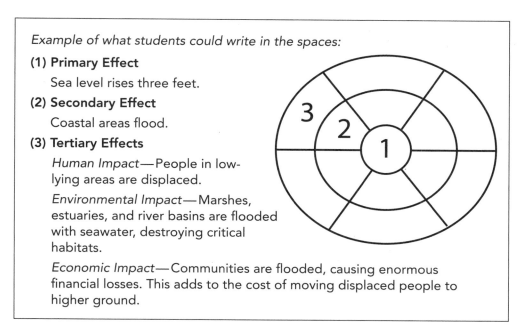

Example of what students could write in the spaces:

(1) Primary Effect

Sea level rises three feet.

(2) Secondary Effect

Coastal areas flood.

(3) Tertiary Effects

Human Impact—People in low-lying areas are displaced.

Environmental Impact—Marshes, estuaries, and river basins are flooded with seawater, destroying critical habitats.

Economic Impact—Communities are flooded, causing enormous financial losses. This adds to the cost of moving displaced people to higher ground.

Figure 7.1: Effects wheel

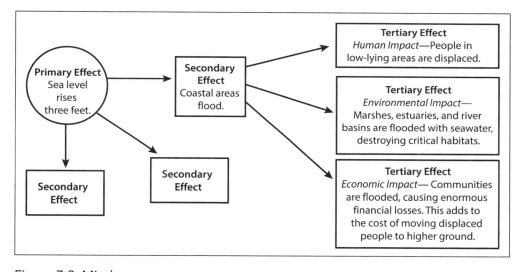

Figure 7.2: Mind map

instruction, these are some models you could use. (Handout 7.1 features a blank effects wheel.) Alternatively, you can choose one model for the whole class to follow, if that better fits your goals.

2. Using these charts or other similar information organizers (e.g., a vertical flowchart), students work in their research groups to choose one effect that they feel is significant (e.g., sea level rises three feet). They will put that effect at the center of their effects wheel, mind map,

or other organizer. (If using a vertical flowchart, the main effect will go at the top of a piece of paper.) They will then list the secondary effects that come from that initial effect (e.g., coastal areas flood). They may list as many secondary effects as they choose. They will split their tertiary effects into three categories:

a. Human impact (on individuals, communities, nations, etc.)

b. Environmental impact (on plants, animals, habitats, ecosystems, etc.)

c. Economic impact (cost to prevent, repair, or mitigate the effect on both the environment and human populations)

  Students should include local effects (if any), national effects, and international effects.

3. Students will most likely need to adapt their charts to reflect the number of secondary effects they choose, but this is an excellent opportunity for them to use critical thinking skills to organize their information in order to communicate it effectively to the class.

Begin!

Climate Change Outcomes: The Ripple Effect

1. Explain that students will work in their research groups to create charts in order to look at the wide range of effects that climate change may have. In particular, they will be addressing the secondary and tertiary effects (the ripple effects) that the initial climate change effect may cause.

2. On the board, provide an example:

Primary Effect:
Sea level rises three feet.
↓
Secondary Effect:
Coastal areas flood.
↓
Tertiary Effects:

↓	↓	↓
Human Impact	Environmental Impact	Economic Impact
People in low-lying areas are displaced.	Marshes, estuaries, and river basins are flooded with seawater, destroying critical habitats.	Communities are flooded, causing enormous financial losses. This adds to the cost of moving displaced people to higher ground.

Ask how the tertiary effects will impact humans, the environment, and the economy. List student ideas on the board under Tertiary Effects.

3. Students should also include at least one international and one national secondary effect. If applicable, they should also include at least one local effect.

4. Point out that students will begin to see these "chain reactions" of effects. This is one reason that it is difficult to predict the consequences of a warming planet.

5. Each team will use the information that they or their classmates found during their research in Session 5 to assist them in completing their charts. They may choose as many secondary effects as they feel they can effectively address. They can also conduct additional research to add to their original information. Students should only use effects that can be supported by research, and they should not stretch the truth just to create a larger chart.

6. Allow about 30–45 minutes for the teams to complete their effects charts. Circulate around the room to help any students who need assistance with understanding the task. When you decide that they have spent enough time working on their charts, regain their attention.

7. Ask each team to reflect on their observations as the results of climate change move from primary to secondary to tertiary effects. Give teams about 10 minutes to decide on two pieces of information they will share with the class. Students will share the following:

   a. The most compelling chain of effects they have in their chart (e.g., a surprising outcome they had not expected)

   b. An effect they feel may be significantly mitigated or eliminated with an action that is being taken now or that will likely be taken in the future

8. Have each team share these two pieces of information with the entire class.

   Note: *It is important that students don't get too caught up in the "doomsday" scenarios. If the conversation is getting too heavy, move on. The next session will be a more uplifting and empowering activity.*

9. See Figure 7.3 (p. 112) for sample student work.

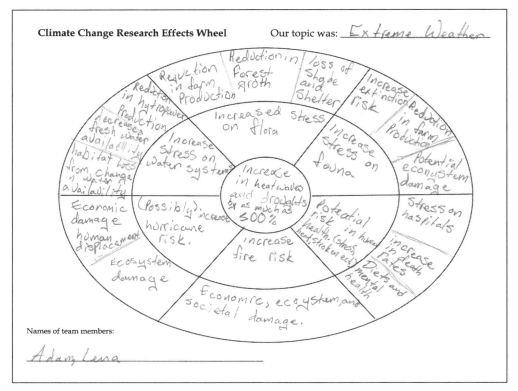

Figure 7.3: Climate change ripple effects (sample student work)

Climate Change Agent Interview

10. Pass out Handout 7.2: Climate Change Agent Interview. The interview features James Balog, a photographer, geologist, and filmmaker who uses STEAM (science, technology, engineering, art, and math) in his work. Give students a few minutes to read the interview and jot down their thoughts. Ask students for their impressions. Take a few comments. Ask, "How has James Balog chosen to tell the story of a changing climate?" Ask students to consider his career path. Ask, "What were some of the reasons he changed direction in pursuit of his passions?" Give students a few minutes to write down notes about this interview and feedback from other classmates.

Reviewing Statements and Questions About Climate Change

1. Ask students to review the list they made in their notebooks about what they had heard about climate change. Ask if they now have enough evidence to determine the accuracy of any more of these ideas. If students have enough evidence to categorize a statement as accurate, they should record it on a sentence strip and post it in the Accepted as Accurate column. If they have enough evidence to categorize a statement as inaccurate, they should record it on a sentence strip and post it in the Accepted as Inaccurate column. If the evidence suggests that an idea on their list is accurate but not definitive, they may post it in the Needs More Information/Evidence/Research column. *In each case, students need to support their decision with evidence.*

2. Now revisit the statements posted during the previous sessions. Ask if anyone thinks a statement should be moved to a different column based on new information they learned during the current session. Students may decide that they need more evidence for a statement that they had previously placed in the Accepted as Accurate category, or they may now have enough evidence to move a statement that had been placed in the Needs More Information/Evidence/Research column to either the Accepted as Accurate or Accepted as Inaccurate column. *In each case, they need to support their decision with evidence.*

3. Ask students if they can write new statements for any of the columns, based on new knowledge gained from this session. *In each case, they need to support their decision with evidence.*

4. Ask students to review the questions that are posted. Are there any questions that they can now answer with their new knowledge? If so, remove the question, write out the answer on a sentence strip, and post this statement in the Accepted as Accurate column. *In each case, they need to support their decision with evidence.*

5. Finally, ask students if they have any new questions to add to the Questions We Have About Climate Change column. Provide a few minutes for discussion and for posting questions.

Assessment Opportunity

You may want to use the assessment probe What Are the Signs of Global Warming? from *Uncovering Student Ideas in Earth and Environmental Science* (Keeley and Tucker 2016) to assess whether students are now able to support the claim that

our planet is warming. Some statements in the probe are direct measurements of a warming planet whereas others are inferences of effects from a warming planet (e.g., sea level rise, ocean acidification, glacial retreat, heavy downpours). This assessment can be found on the Extras page: *www.nsta.org/climatechange*.

Background for Teachers

Pedagogy

This session is an exercise in critical thinking. Students are challenged to think clearly and rationally about a specific problem. They are also asked to demonstrate an understanding of logical connections between ideas. They do this by identifying secondary and tertiary effects and then separating the tertiary effects into human, environmental, and economic impacts. Do not be surprised if students notice overlap and connections among these three categories. An environmental impact may be decreased pH of water, which means that oyster larvae and other marine invertebrates are unable to form shells. This will cause an increase in production costs for cultivating oysters. It can also cause reduced ocean productivity, which will lead to a shortage of food for people. This overlap and connection of issues will provide a good source for student discussion as they attempt to sort out these third-level effects.

As students share their effects wheel with the class, they are asked to speculate on a chain of events that can be solved by human action. This prepares students for the climate change solutions they will research in Session 8.

Reference

Keeley, P., and L. Tucker. 2016. What are the signs of global warming? In *Uncovering student ideas in Earth and environmental science*, 77–81. Arlington, VA: NSTA Press.

SESSION 8

Climate Change Solutions

Introduction

This session is about hope. After students have learned a great deal about the serious implications of a warming planet, it is important for them to have reasons to be hopeful and to see solutions that are being implemented and planned for the sustainability of our planet and their future.

Students form new research teams, bringing important information about their research topics with them to this new group. Students choose a nonprofit group, governmental agency, or corporation at the local, regional, state, national, or international level that is working to solve the climate crisis in a variety of ways. Teams choose a particular action that their organization is taking to mitigate or adapt to climate change. They use similar methods from Session 7 to chart the ripple effects of this positive change on secondary and tertiary levels. Finally, they choose the most compelling chain of effects they would like to highlight and share with the class. The presentations paint a powerful positive vision of what their future could look like and provide inspiration to take a wide range of actions.

There are two Climate Change Agent interviews in this session. One features Students for Sustainability, a student-led group whose members are working to reduce the carbon footprint of their school, their community, and their region. This group serves as an example of actions that are being taken at the student level, and it will encourage your students to think of ways in which they can take action. The other interview features Dr. Ziv Hameiri, who is working on cutting-edge solar energy technology that may transform how we capture the Sun's energy.

By combining inspirational ideas from a wide range of groups, agencies, and organizations, students may be motivated to find solutions of their own. Session 9 will provide opportunities to put those solutions into action.

Objectives

1. To give students opportunities to research what a range of organizations—from local entities to international groups—are doing to mitigate the effects of climate change

2. To give students a broader picture of what the world can do to effectively address the issues of global warming and climate change

3. To integrate the knowledge gained in Sessions 5 and 6 and apply it to consequential thinking of climate change solutions and their positive effects

4. To learn what the members of one student-led group are doing in their school and community to reduce their carbon footprint

What You Need

Gather the following materials.

For the class:

- ❑ The current list of questions and statements on the wall OR access to the electronic documents with questions and statements, for revision

- ❑ Sentence strips for additional statements (if you're using the wall columns)

- ❑ Marking pens (if you're using the wall columns)

For each group of four students:

- ❑ Laptop or electronic device with internet access

For each student:

- ❑ Science notebook

- ❑ Handout 8.1: Climate Change Solutions Effects Wheel or other chart design you provide (optional)

- ❑ Handout 8.2: Climate Change Agent Interview

- ❑ Handout 8.3: Climate Change Agent Interview

    Note: All handouts are located on the Extras page: www.nsta.org/climatechange.

Preparation

Several Days Before the Class

1. Research helpful websites of special-interest groups that represent your region and are working on solutions to the climate crisis. Make sure there is practical information about what the group is doing to promote a reduction in greenhouse gases—not just policy statements. (Optional: Contact individuals in local government, corporate, and nonprofit groups who are working on climate change and can serve as resources for students.)

2. Provide resources and suggestions for students on how to collect the data needed to calculate their carbon footprint. (For instance, students should look at electric bills, add up how many flights they've taken in the last year, calculate the volume of trash produced weekly by their families, etc.) Students won't be using this information in Session 8. Instead, they will need the data for the carbon footprint calculator used in Session 9 (pp. 126–129). However, they should start gathering the data now in order to have that information ready by the time Session 9 begins.

Suggestions for different carbon calculators are provided in the Preparation section of Session 9. If your students are going to use the Stanford University calculator, they can access an online chart that will help them gather the data needed for the calculator. Students can register and save their data, using it for comparison as they explore what changes they can make to lower their carbon footprint. You can access the chart here: *http://footprint.stanford.edu/documents/CalculatorPrep.pdf.* The chart can also be found on the Extras page (*www.nsta.org/climatechange*).

The Day Before the Class

Decide how you will reorganize students into new teams. Each team will be made up of students from different research topics in order to provide a variety of information and perspectives while they research solutions to climate change.

Begin!

Climate Change Solutions: Near and Far

1. Explain that students will be creating new teams to complete charts of solutions to climate change.

2. Once the new teams are formed, have students look through their notebooks and see if they can think of a particular organization or type of organization that is creating solutions to the climate crisis.

3. Have each team choose an organization to research in order to find out what it is doing to solve the climate crisis. If any teams are having trouble choosing an organization or agency from their notes, provide the following list to help them come to a decision:

   - Students in their grade level, both in the United States and around the globe

   - Local politicians (city council, mayor, county commissioners, etc.)

   - Local, regional, state, national, or international

     - nonprofit organizations,

     - governmental agencies,

     - corporations, or

     - universities and research institutes.

   Guide students toward their decision using this list and the prior research done on special-interest groups in your region.

4. Explain that teams will take on the perspective of their chosen organization as they brainstorm a list of actions that are being done or could be done by their group in order to mitigate or reduce the negative effects of climate change.

5. Provide students with an example to show the ripple effects of one action:

   High school students eat more locally produced food.

   Vehicles travel less to bring goods to market.

   Less CO_2 emitted into the atmosphere.

6. Teams will now investigate what their chosen groups or organizations are doing to reduce the effects of global warming and climate change. They will make a list in their notebooks of what they discover. You may need to assist them in their research, directing them to solutions that reflect the interests of their groups or organizations.

7. After gathering a substantial amount of information, teams will use their research to help choose a specific action that their groups or organizations are doing to make significant strides in mitigating or reducing climate change. They may also choose an action that they think the group is capable of doing but currently not doing.

   Note: Students may discover that local groups do not have the same degree of influence as national or international groups. That is fine. It will be interesting for students to see the wide range of actions and effects from local to international organizations, which will also help them decide how best to use their individual energy and interests if they want to effect change on a warming planet.

Climate Change Solutions: Results of the Ripple Effect

1. Once all teams have chosen their actions, they will create an effects wheel, mind map, or other method of organizing their information. (If you choose, you can distribute Handout 8.1, which features a blank effects wheel.) They will need to find secondary and tertiary effects that come from the initial action and add those to the chart.

2. Each team will use the information from their research to list as many secondary and tertiary effects as they can. They should use only reasonable effects.

3. Allow about 25–45 minutes for the teams to complete their charts. When you decide they have added enough information to their charts, draw their attention.

4. Ask teams to reflect on what they have learned about their groups or organizations. They should note the results of the primary action as the ripple effect goes out to the secondary and tertiary levels. Ask them to take about 10 minutes to decide what is the most compelling chain of effects they have in their chart. Allow students to define what they think *compelling* means.

5. Ask each team to share their most compelling chain of effects with the class. Allow time for comments or additional ideas.

6. Once all of the teams have shared their ideas, conduct a discussion on the relative importance of the actions they have learned about in

this session. Students may note that larger organizations may have a greater effect on reducing carbon emissions. Other students may note that smaller groups can provide individuals with a chance to make a difference. Encourage students to consider all types of actions and how they add up.

7. Have students take some time to add their thoughts to their notebooks.

8. See Figure 8.1 for sample student work.

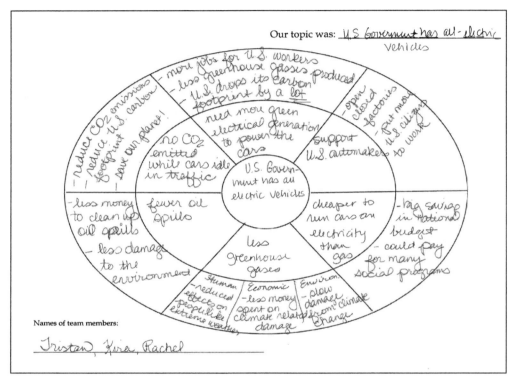

Figure 8.1: Ripple effects of climate change solutions (sample student work)

Climate Change Agent Interviews

1. Distribute to each student a copy of Handout 8.2: Climate Change Agent Interview. This interview features Students for Sustainability, a student-led high school group whose members are working to lower the carbon footprint at their school, in their community, and across their state. Allow a few minutes for students to read the interview and make comments in their notebooks.

2. Ask for students' impressions of this interview. Take as many answers as time allows. Ask, "What actions did this student group take that you would like to try?" Point out that young people across the country and around the world are working in a variety of ways to make the planet more sustainable. Explain that in the next session, students will use the data they have been collecting to calculate their own carbon footprint and come up with ideas for ways they can reduce it.

3. In Session 9, you may choose to have students come up with actions they can take as a class, school, or club. If you are planning to do this, mention to students that they will be coming up with these action plans in the next session. Have students jot down in their notebooks ideas for starting projects in their school or community.

4. Distribute to each student a copy of Handout 8.3: Climate Change Agent Interview. The interview features Dr. Ziv Hameiri, an electrical and computer engineer working on cutting-edge solar energy technology that may transform how we capture the Sun's energy. Allow a few minutes for students to read the interview and make comments in their notebooks.

5. Ask for students' impressions of this interview. Take as many answers as time allows. Ask, "How is Dr. Hameiri making a difference to solve the climate crisis? Will his efforts have local, national, or global effects?"

Reviewing Statements and Questions About Climate Change

1. Ask students to review the list they made in their notebooks about what they had heard about climate change. Ask if they now have enough evidence to determine the accuracy of any more of these ideas. If students have enough evidence to categorize a statement as accurate, they should record it on a sentence strip and post it in the Accepted as Accurate column. If they have enough evidence to categorize a statement as inaccurate, they should record it on a sentence strip and post it in the Accepted as Inaccurate column. If the evidence suggests that an idea on their list is accurate but not definitive, they may post it in the Needs More Information/Evidence/Research column. *In each case, students need to support their decision with evidence.*

2. Revisit the statements posted during the previous sessions. Ask if anyone thinks a statement should be moved to a different column based on new information they learned during the current session.

Students may decide that they need more evidence for a statement that they had previously placed in the Accepted as Accurate category, or they may now have enough evidence to move a statement that had been placed in the Needs More Information/Evidence/Research column to either the Accepted as Accurate or Accepted as Inaccurate column. *In each case, they need to support their decision with evidence.*

3. Ask students if they can write new statements for any of the columns, based on new knowledge gained from this session. *In each case, they need to support their decision with evidence.*

4. Ask students to review the questions that are posted. Are there any questions that they can now answer with their new knowledge? If so, remove the question, write out the answer on a sentence strip, and post this statement in the Accepted as Accurate column. *In each case, they need to support their decision with evidence.*

5. Finally, ask students if they have any new questions to add to the Questions We Have About Climate Change column. Provide a few minutes for discussion and for posting questions.

Resource

Project Drawdown (*www.drawdown.org*) is currently the most comprehensive plan ever proposed to reverse global warming. Leaders of the project worked with a diverse group of researchers from around the world to identify, research, and model the 100 most substantive existing solutions to address climate change. Their efforts resulted in a blueprint for rolling back global warming within 30 years. Project Drawdown shows that humanity already has the means to reverse global warming. The solutions are in place and in action. The following includes a summary of the 100 solutions by rank: *www.drawdown.org/solutions.* Created in conjunction with the project is the book *Drawdown,* edited by environmentalist and entrepreneur Paul Hawken.

Background for Teachers

Pedagogy

Rather than give students a list of organizations or agencies to consider at the beginning of this session, they are asked to brainstorm a list based on their experiences throughout the entire unit. This is to engage students in thinking about who may be working to solve the climate crisis and to underscore that a lot of

work is being done at a number of levels. The Climate Change Agent interviews can also provide ideas for students to draw from.

After identifying the organization, agency, or group they want to research, students are asked to be critical thinkers and identify a solution to the climate crisis connected to their chosen group.

In Session 7, students created a diagram that showed the ripple effects of specific aspects of climate change. In this session, they are challenged to identify the ripple effects of a specific solution. As in Session 7, they will need to apply knowledge from all of the lines of evidence reported on in Session 6 as they think about the ramifications of one discrete solution.

For years in education, many taught that science was a linear process: Ask a question, form a hypothesis, design an experiment, gather data, and draw conclusions. This does not reflect the real world of science. Science relies on creativity and thinking outside the box. It is an iterative process where conclusions fuel new questions and evidence is used to design solutions. This is reflected in the science and engineering practices of the *NGSS,* and it is the focus of this session.

References

Hawken, P., ed. 2017. *Drawdown: The most comprehensive plan ever proposed to reverse global warming.* New York: Penguin Books.

NGSS Lead States. 2013. *Next Generation Science Standards: For states, by states.* Washington, DC: National Academies Press. *www.nextgenscience.org/next-generation-science-standards.*

SESSION 9

Connecting to Your Community

Introduction

Now that students have a broad picture of what their future may hold as the planet warms, it's important to provide them with concrete actions they can take to make a difference. Session 9 begins with students calculating their own carbon footprint and comparing it to that of the average person in the United States. Once they have an idea of how much CO_2 they generate in a year, students come up with changes they can make at home and calculate the amount of CO_2 saved.

Students then read a Climate Change Agent interview about another student-led sustainability club, Schools Under 2°C. They compare and contrast the work done by Students for Sustainability (in Session 8) to Schools Under 2°C. Using information from the interviews as a launch pad, a brainstorm activity ensues with students creating an action plan for their school.

The session culminates with students planning actions they can take in their communities. Teachers may work with students to create a school sustainability club or other organization to implement their plans and work toward a brighter future for all.

Objectives

1. To have students calculate their own carbon footprint and compare it to the average person in the United States

2. To provide students with opportunities to get involved in reducing their personal carbon footprint as well as that of their families, their school, and their community

3. To provide students with concrete ways they can personally make a difference in affecting positive outcomes in the area of climate change

4. To have students use a computer simulation to model the impact of proposed changes intended to lower carbon footprints at home and at school

What You Need

Gather the following materials.

For each student:

☐ Science notebook

☐ Access to the internet to complete an online carbon footprint calculation

☐ Copy of Handout 9.1: Climate Change Agent Interview

☐ Copy of Handout 8.2: Climate Change Agent Interview (from Session 8)

☐ Copy of Handout 9.2: Carbon Footprint Reduction Pledge

Note: All handouts are found on the Extras page: www.nsta.org/climatechange.

Preparation

One Week Before the Session Begins

Explore the following suggested carbon footprint calculators to determine which one will be best for your students to use in class. Some calculators are more comprehensive and will take longer to complete. Gauge the time necessary to complete your selected calculator in order to plan your instructional time.

Recommended Calculators

- Stanford University has created a student-centered carbon calculator that adjusts for the types of activities that are most common among young people. It measures home energy and appliances, food, personal purchases, and transportation. You may choose a version suited for middle or high school students. Each student's carbon footprint is compared to that of the average person in their state and in the world. Students are able to compete with other students around the world, as well. There is also a version available in Spanish. You can find the calculator here: *https://i2sea.stanford.edu/calculate.* It takes approximately 25–30 minutes to calculate a carbon footprint using this calculator.

  A chart is provided for students to gather the data needed for the calculator *(http://footprint.stanford.edu/documents/CalculatorPrep.pdf).* Students can register and save their data and use them for comparison as they explore changes they can make to lower their carbon footprint.

Other Calculators

- University of California, Berkeley's CoolClimate Calculator

  This calculator has both quick and more in-depth options while providing ideas for lowering your carbon footprint.

  https://coolclimate.berkeley.edu/calculator

- Carbonfootprint.com

  This widely used calculator compares students' footprints to those in their country and around the world. It promotes the idea of offsetting your carbon footprint by selecting ways to sequester carbon, such as tree planting.

  www.carbonfootprint.com/calculator.aspx

- The Nature Conservancy

  After completing your carbon footprint calculations, this web page provides suggestions for ways to reduce your carbon footprint in the areas of transportation, home energy, and shopping.

  www.nature.org/en-us/get-involved/how-to-help/consider-your-impact/carbon-calculator

  Note: *There are small to large variations in each carbon calculator, causing different results despite inputting the same data. This is due to the calculators using different metrics to obtain their results. It is suggested that you choose one calculator for the class in order to keep the results consistent for comparison between students.*

A Few Days Before the Class

1. Have students gather the data they will need to input into the carbon footprint calculator. If you use the recommended calculator from Stanford, a chart for this data has been created for you. You can gather necessary data using this chart or using a similar chart found on the Extras page: *www.nsta.org/climatechange*.

2. Consider how you will allow students to begin planning their actions/projects/plans for sustainable changes at their school. Adult mentors, school advisors, or club leaders may need to be involved to support the students. Students may also launch their ideas all on their own.

Begin!

Calculating Your Carbon Footprint

1. Have students think about how their carbon footprint might compare to that of the average person in the United States. Ask, "Why do you think this?" Suggest that they consider their home energy use, shopping habits, transportation, and the like. Have them write their justification for their estimates in their notebooks.

2. Have students look up the carbon footprint of the average person in the United States. (This information was provided for them in Session 3. See Handout 3.2: International CO_2 Levels in 2015.)

3. Provide students with the link to your selected carbon footprint calculator and have them begin their calculations. Encourage students to work together to help answer any questions they may have about the metrics required for the calculator. This will help them improve their problem-solving skills. If students cannot get answers from their classmates, have them raise their hands to request your assistance. The calculations may take up to 30 minutes to complete.

4. Once students have completed their calculations, have them jot down in their notebooks any questions they have or any points they wonder about. Ask them to revisit their prediction for how their footprint compares with that of the average person in the United States. Have them evaluate why their prediction was the same as or different from their calculated footprint. Have students compare their footprint to that of people in other countries (listed on Handout 3.2: International CO_2 Levels in 2015) and note which countries have citizens with similar footprints.

5. In small groups, have students share their results and discuss what activities or factors cause similarities or differences in their footprints.

6. Allow for lively discussion, and then gather the attention of the class. Ask what insights they now have about their own carbon footprints. Take as many comments as time allows.

7. Have students revisit the specific data used to calculate their personal carbon footprint and make realistic suggestions in their notebooks about what they could do to lower their footprint. Have students calculate the amount of CO_2 saved by these actions.

8. Have students input this new data into the calculator to see how their footprint changes. Based on this new information, have each student create a plan for how they will lower the carbon footprint in the next few weeks or months. Ask students to record their plans in an easily accessible place (e.g., a notebook) so they can regularly refer to them and monitor their progress over the next several months.

Take the Challenge!

1. Distribute Handout 9.2: Carbon Footprint Reduction Pledge to each student.

2. Have students think of a few actions they promise to continue, to add, and to stop in order to make this pledge:

   I pledge allegiance to the Earth and all the life that it supports. One planet in our care, irreplaceable, with sustenance and respect for all.

3. Suggest that students post their pledge in a place where they can review it and revise it over time.

4. If possible, provide some time each week for students to check in with their plans for lowering their carbon footprint and share successes, difficulties, and insights.

Climate Change Agent Interview

1. Distribute to each student Handout 9.1: Climate Change Agent Interview, which features Rayan Krishnan from Schools Under 2°C. Allow a few minutes for students to read the interview. Have students jot down notes about their impressions of the interview.

2. Ask, "What interested you most about Rayan Krishnan?" Have students talk with others at their tables. After a few minutes, ask for students to share their thoughts.

3. Have students review Handout 8.2: Climate Change Agent Interview from Session 8. (The interview features Students for Sustainability.) Ask students to work in small groups to compare and contrast the focus and activities from the two student groups and jot down any ideas they think might work at their school and in their community. Allow time for a thorough discussion.

Note: You may not have the time or ability to support students in taking on sustainability projects at their school or in their community. If that is the case, skip to the last part of this session, Reviewing Statements and Questions About Climate Change (p. 132).

Creating Change at Your School and in Your Community

1. Have students fill in the blank of the following statement: "If I were in charge of this school or my community, I would _____ to make the school/community more sustainable and lower its carbon footprint." They can name as many items as they want in the time you give them to complete the activity.

2. Working alone, have students make a list of possible changes and programs that could be implemented. They can include the ideas they came up with during the previous fill-in-the-blank activity above and the ideas inspired by the two Climate Change Agent interviews they just read.

3. Have students share their ideas with others at their tables, with each student speaking while the others listen without making comments. Once each student has shared their ideas, have students work together to make one list of ideas for each table group.

4. Have a spokesperson for each table share the group's list of ideas with the rest of the class. The spokespeople don't have to go into great detail at this time; they should just give enough information so the class gets the idea of the activity, project, or plan (APP). Record their ideas on the board or project them from an electronic device so all can see.

5. After each idea is presented, have table groups rank the ideas using the following criteria. They can further define the parameters if you want. However, it is fine to leave them a bit vague for now.

   • **Viability:** What is the possibility that we can do this?

      1 - difficult   2 - not too difficult   3 - easy

- **Student interest:** Are we interested in doing this APP?

  1 - not really interested   2 - moderately interested   3 - we like it

- **Cost:** How much will it cost to implement the APP?

  1 - high cost   2 - moderately expensive   3 - inexpensive or free

- **Time:** How long will it take to complete the APP?

  1 - a long time   2 - not too long   3 - we can do this quickly

6. Instruct table groups to add up the four values they assigned to each idea presented in order to come up with a total number for each APP.

7. Return to the idea list generated on the board or on the electronic device. If using a board, have one student from each table bring up the group's total numbers for each APP and list them on the board beside the corresponding idea. If projecting the list on an electronic device, either you or one student from each table can log the numbers into the device.

8. Give the class a few minutes to evaluate the list. Ask if they have anything to add or clarify. Sort the list again, ranking the APPs from the highest (12) to the lowest (4).

9. Conduct a class discussion about what APPs the students would like to initiate. Be sure to assess those that may rank high (easy to do, low cost, and short timeline) but don't generate much student interest, and place them lower on the list.

10. Once a short list of APPs with the highest interest and most potential for success has been created, ask students to identify themselves as leaders or part of the support team for each APP. Provide guidance on realistic goals and numbers of APPs that can be taken on at once. Remind students that once they finish one project, they can always start another from the list or come up with new ones they want to do.

11. Have students regroup based on the APPs they are interested in working on. Ask them to begin planning how they might address their APPs. Students should list what resources they need, who they need to meet with, what obstacles they currently see, etc. Ask how they will measure the success of their project. Have them include that in their plans. They can design a timeline and insert items from the lists they made. What is the environmental impact their APP will have? If possible, students can try calculating the pounds of CO_2 that will be prevented from entering our atmosphere as a result of their APP.

12. Circulate around the room to make sure each student is involved. It is fine if a student wants to take on an APP alone. Answer questions,

but try to provide as little guidance as possible so students can take responsibility for their ideas and APPs. The goal is not to succeed with ease but to learn how to make realistic plans and adjust the plans as they unfold to address any setbacks, bureaucratic obstacles, or other unforeseen issues that may arise.

13. At this point, the projects will take on a life of their own. Students may want to start a sustainability club, combine efforts with a club already at the school, partner with a community group, and more. Share your plans for scaffolding and support as students begin planning their APPs so that engaged and excited students can move their ideas forward. This may require time outside of class as you move on to a new unit. Some students may effectively address their ideas all on their own. Others may need the support of adult mentors, school advisors, or club leaders to launch their projects. Provide contact information for individuals who can serve as supporters for the students.

Reviewing Statements and Questions About Climate Change

1. Ask students to review the lists in their notebooks on what they had heard about climate change. Pair students and have them share items on their lists that they have yet to classify as accurate or inaccurate. Ask, "What evidence do you still need to determine if these statements are accurate or inaccurate? What would you need to do to get that evidence?"

2. Now revisit the statements posted during the previous sessions. Ask if anyone thinks a statement should be moved to a different column based on new information they learned during the current session. *In each case, they need to support their decision with evidence.*

3. Ask students if they can write new statements for any of the columns, based on new knowledge gained from this session. *In each case, they need to support their decision with evidence.*

4. Ask students to take five minutes to write about the most important concepts they have learned since the beginning of the unit. Once all have finished, ask volunteers to share the most important concepts they learned during the unit.

5. Next, ask students to review all of the questions that have not been answered. What information or evidence do they think they need to answer the remaining questions? This is a great time to encourage

students to explore careers in science, technology, engineering, and math. People in these fields hold the keys to our future.

6. Ask students to return to their initial thoughts on the people that they met in the networking activity in Session 1. In what way was each of these people a change agent? Ask students to add their own name to the list of important people. Ask, "What can *you* do to be a change agent?"

Looking to the Future

1. Ask, "What do you hope your city/town, state, nation, or world will look like in 5 years? In 10 years? In 50? In 100?" Have students discuss what they hope the future looks like at each milestone year. What is working? What is still unsustainable? Have students describe their visions of the future in their notebooks. They can use words and drawings.

2. Congratulate all students for their hard work, great ideas, and dedication to making their world a better place.

Extending the Session

1. Students may want to join carbon footprint competitions or create their own, either as teams or as individuals. They might want to have competitions between classes or between schools.

   Two established carbon footprint competitions are listed here:

   a. Taming Bigfoot: This competition was initially designed by the Local 20/20 Climate Action Outreach group in Port Townsend, Washington. Participants can download the Taming Bigfoot app to their smartphones (it works with Android and iOS platforms) and input their data each month. Find out more here: *www.taming-bigfoot.org.*

   b. Stanford University's carbon calculator provides an opportunity for your students to compete with schools around the world: *http://web. stanford.edu/group/inquiry2insight/cgi-bin/i2sea-r3b/i2s.php ?page=iscfc.*

2. Students may be interested to learn more about the lifestyles of people in countries that have a similar carbon footprint to that of the United States. Students can research countries that have smaller footprints and learn what they do to put out less CO_2.

3. Students can imagine what the future will look like in 5, 10, and 50 years from the perspective of each Climate Change Agent. How will

these people be involved in climate change mitigation and adaptation? How will their work affect the planet in those milestone years?

Background for Teachers

Use of Standards

To reinforce the idea that small changes in one's lifestyle can have an impact, consider presenting the crosscutting concept of scaling. What is the cumulative effect of all of the changes made by the students? What if everyone in the school made similar changes? Expand this scaling to the students' community, state, country, and beyond. The students are ready now to take on the world!

Be sure a new student group starts with a focused project. The *NGSS* provides guidance for this work. As students plan, have them identify criteria for success, realistic constraints, and opportunities for revision and refinement of their plans. These standards were designed to prepare students for real-world problem solving. Put them to work!

Pedagogy

"Learning [is] a *process* that leads to *change,* which occurs as a result of *experience* and increases the potential for improved performance and future learning."

—Ambrose et al. 2010 (p. 3)

This lesson begins with a simple question: How does your carbon footprint compare with that of the average U.S. citizen? Use this to first assess if students understand the term *carbon footprint.* In this case, *footprint* is a metaphor for the total impact that something has. *Carbon* is shorthand for all of the various greenhouse gases that contribute to global warming. Therefore, an individual's carbon footprint is the best estimate of how much greenhouse gas that individual contributes directly or indirectly to the atmosphere. It is usually expressed in equivalent tons of carbon dioxide (Berners-Lee and Clark 2010).

The second purpose of this leading question is to prompt students to access and apply information from the graphs and data tables they have reviewed. They will find this information in Handout 3.2: International CO_2 Levels in 2015.

The third purpose is to prompt your students to think about their own activities and the energy sources on which they depend. In Session 8, students considered the impact of solutions made by groups, agencies, or organizations. In this unit, students reflect on and commit to making personal changes.

The hallmark of learning is that it changes thinking in a way that changes behavior. Throughout this unit of study, students have learned to understand

how to evaluate the validity of evidence and how to use this evidence to inform their own thinking on the topic of climate change. The purpose is to give students the tools to build an understanding of complex topics based on evidence rather than opinion.

Analyzing the data and considering the ramifications of climate change can be overwhelming and discouraging. In Session 8, students worked through the ripple effects of changes that can be made by groups, agencies, or organizations. To many students, that can still seem pretty distant from their personal reality.

In Session 9, students run simulations to determine what kinds of changes in their daily lives, both small and large, can add up to make a difference.

Assessment Opportunities

Student notebooks should be used as an assessment tool and as a final evaluation of student understanding and progress for the entire unit.

Resources

1. The website for Schools Under 2°C (*www.schoolsunder2c.org*) has ideas and tools that can be used by students to reduce their carbon footprints. Schools can also sign a pledge to reduce their carbon footprint and join schools from across the country and around the world!

2. Started in 2007 by 9-year-old Felix Finkbeiner of Germany, Plant for the Planet now has student climate ambassadors on six continents! Felix was inspired by Wangari Maathai of Kenya, the Nobel Peace Prize winner who planted 30 million trees in 30 years. Felix decided that children could help plant a million trees in every country to sequester carbon and create a livable planet. Working with the United Nations Environmental Programme, Plant for the Planet has been granted the role of counting trees planted by students worldwide. Find out more here: *www.plant-for-the-planet.org*.

3. The EPA's President's Environmental Youth Award (PEYA) recognizes outstanding environmental projects by K–12 youth. Each year, the PEYA program honors a wide variety of projects developed by young individuals, school classes (kindergarten through high school), summer camps, public-interest groups, and youth organizations to promote environmental awareness. Thousands of young people from all 50 states and the U.S. territories have submitted projects to EPA for consideration. Winning projects in the past have covered a wide range of subject areas, including the following:

- Restoring native habitats
- Recycling in schools and communities
- Construction of nature preserves
- Tree planting
- Installing renewable energy projects
- Creating videos, skits, and newsletters that focused on environmental issues
- Participating in many other creative sustainability efforts

    Find out more about the PEYA program here: *www.epa.gov/education/presidents-environmental-youth-award*.

4. Check out the many excellent resources on climate change from the National Wildlife Federation (NWF), including NWF's Climate Classroom website (*https://climateclassroom.org*). Investigate what other schools and organizations are doing to educate their communities and take action on climate change. You can also visit the following web page for tips on how to tackle climate change: *www.nwf.org/Eco-Schools-USA/Become-an-Eco-School/Pathways/Climate-Change/Tips*.

5. Our Climate/Our Future (*https://ourclimateourfuture.org*) provides award-winning climate education resources.

6. Earth911 (*https://earth911.com*) strives to encourage consumers to reduce, reuse, and recycle.

7. The Climate Reality Leadership Corps (*www.climaterealityproject.org/leadership-corps*) is a global network of activists committed to spreading awareness of the climate crisis and working for solutions to the greatest challenge of our time. The program provides training in climate science, communications, and organizing in order to better tell the story of climate change and inspire communities everywhere to take action.

8. Teachers may want to join NOAA's Planet Stewards Education Project. It provides formal and informal educators with the knowledge and resources to build scientifically literate individuals and communities that are prepared to respond to environmental challenges.

References

Ambrose, S., et al. 2010. *How learning works: Seven research-based principles for smart teaching*. San Francisco: Jossey-Bass.

Berners-Lee, M., and D. Clark. *The Guardian*. 2010. What is a carbon footprint? June 4. *www.theguardian.com/environment/blog/2010/jun/04/carbon-footprint-definition*.

Index

Note: Page references in **boldface** indicate information contained in figures or tables.